Adapting Buildings for Changing Uses

Adapting Buildings for Changing Uses

Guidelines for change of use refurbishment

David Kincaid

LONDON AND NEW YORK

First published 2002 by Spon Press
11 New Fetter Lane, London EC4P 4EE

Simultaneously published in the USA and Canada
by Spon Press
29 West 35th Street, New York, NY 10001

Spon Press is an imprint of the Taylor & Francis Group

Typeset in Sabon by
Florence Production Ltd, Stoodleigh, Devon EX16 9PN
Printed and bound in Great Britain by
TJ International Ltd, Padstow, Cornwall

British Library Cataloguing in Publication Data
A catalogue record for this book is available from the British Library

Library of Congress Cataloging in Publication Data
Kincaid, David, 1935–
Adapting buildings for changing uses: guidelines for change of use refurbishment/
David Kincaid.
p.cm.
ISBN 0-419-23570-1 (pbk.: alk. paper)
1. Buildings–Remodeling for other use. I. Title.

TH3401.K56 2002
720′.28′6–dc21 2002020494

ISBN 0–419–23570–1

Contents

Figures and tables

Figures

Tables

Preface

Most of the information and many of the ideas incorporated in this book are drawn from research at University College London (UCL) which focused on change of use adaptations occurring in Greater London in the mid-1990s. At that time there was an evident surplus of office space alongside what has by now become an almost chronic shortage of housing. This created a clear opportunity to examine the mechanisms of decision-making and implementation involved in converting redundant offices to housing, an activity of particular interest to policy-makers at the time. However, it seemed evident that to understand these mechanisms for one type of change it would be important to look also at all other types of change; otherwise the decisions examined would be limited to those taken within a project and the investigation of implementation might have amounted to a simple description of the activity. This wider view also provided an opportunity to gain some insight into related problems of urban blight created by the evident surplus of derelict industrial buildings in London at that time. This proved to be very relevant to the question of conversion to housing as found by Lord Rogers' Urban Taskforce[1] (Department of the Environment, Transport and the Regions, 1998) several years later. Perhaps then the book can make some contribution to the solution of these urban problems even though no claim is made to have special insight into the causes.

The broad scope of the research led to the possibility of providing new insights, based on extensive data, into the problems and opportunities of what in Canada and the USA was being called 'adaptive reuse'. London was felt to be sufficiently large and complex to be the source of a wide-ranging set of building data at any time. Also, political and economic circumstances had by the mid-1990s become particularly dynamic and new flexibilities in relation to planning regulation seemed to be emerging: promising circumstances for research, and possibly even more promising for the production of a set of guidelines for those faced with the problems and opportunities posed by adaptive reuse. This book came into even clearer focus as an objective when it became apparent during the research that the available literature on the subject provided little guidance on what adaptations might be most appropriate or how best to proceed except in relation to costings and business cases. In fact, most of the literature was in the form of picture books showing examples of conversions and describing 'before and after' with no commentary on issues of choice, process or success. Thus the research and this subsequent book emerged from an opportunity to investigate a very old and established aspect of the built environment from a perspective that had hitherto been largely ignored.

In writing a book on a topic which is a daily activity for many designers and contractors I am very conscious of the need to avoid claiming an insight into to what is already well known or mistaking a unique circumstance for a general principle. However, the exposure the material in the book has had to very experienced practitioners among the Industrial Partners to the research and the wide range of people we interviewed encourages me to think that this is not a major hazard. In any event, the book should at least provide a serious analysis of current practice in the UK and at best present some new approaches to decision-making and implementation in a key sector of the property industry.

Acknowledgements

My thanks must first go to the sponsors of the research on which this book is based. It was through the Department of the Environment's sponsorship of the 'Construction Maintenance and Refurbishment Research' programme and the support of seven 'Industrial Partners' that the research team was able to investigate refurbishment activities in London between 1994 and 1996. These guidelines are based on this work, and I would like to thank the DoE and its successor the DTI, BT Group Property, Bernard Williams Associates, the British Institute of Facility Management, DEGW Ltd, Weatherall Green & Smith, Willmott Dixon and YRM plc for their support. They provided not only the financial resources necessary to this endeavour, but the insights, data and access essential to gain at least some understanding of this complex activity.

The author and the research team are grateful to all the many people who helped in the research but would like to acknowledge particularly those people who made up the steering group for the research and guided and encouraged this publication. The steering group were:

- Alan White and Ed Costelloe of BT Group Property
- John Desmond of Bernard Williams Associates
- John Crawshaw of the British Institute of Facility Managers
- David Chippendale of Chippendale Consulting and Research
- David Tong and Dr Frank Duffy of DEGW Ltd
- Ian Dodwell of Weatherall Green & Smith
- Jim Murray of Willmott Dixon
- Peter Hammond and Alan Bacon of YRM plc (formerly)
- Peter Pullar-Strecker, Richard Rooley and Michael Ridley of the CMR LINK Programme Committee
- Neil Jarrett of the DoE
- Jacqui Williams of the EPSRC.

Particular thanks also to the many developers, architects, engineers, surveyors, contractors, investors, regulators and clients and their staffs without whose time and patience we would not have been able to carry out the research or produce these guidelines.

Thanks also to Ghada Madfai for her patient work in 'scoring' economic uses for the comparator and Charles Egbu for his contribution to the team's understanding of selective demolition.

Finally, the author wishes to acknowledge the major contribution of my research colleagues Professor Bev Nutt and Peter McLennan to many parts of the book. Both continue to investigate and teach on the subject of this book and related areas of Facilities Management at UCL.

Chapter 1

Adaptive reuse

1.1 Introduction

In considering the potential for buildings to be adapted to different uses it is helpful to start by looking at the question from the simple two-dimensional standpoint of the floor area required for different specific activities. This ignores height, strength and many other ultimately important physical characteristics that are necessary to detailed design, but by focusing on activity, not 'use category', allows us a clearer view on the commonalities of human activities whatever the use setting. This two-dimensional space view was explored in research done in the 1960s which looked at the problems created by rapid growth and change in buildings designed for hospital and school use. Peter Cowan, then at UCL, investigated this question and reported his findings to the Bartlett Society[2] for a large and varied number of different activities (Cowan, 1963). He showed that when all sizes of spaces used for a generic set of human activities were plotted against frequency of occurrence, the peak of the curve of space provision occurred at only 20 square metres and fell away sharply thereafter as space size increased. Also, spaces as small as 2.5 square metres were found to be appropriate for a wide range of useful and necessary activities. These findings indicated that the potential for buildings which may appear constrained by internal configurations, shape or structure (as well as for those that are large and relatively unconstrained physically) to be adapted for a wide variety of uses is not especially limited by the space needs of a significant range of human activities.

This suggests that most buildings are physically suitable for adaptation to most uses, and influenced the proposition that 'long life – loose fit', which was popular in the 1960s, should be a guiding principle behind most design briefs. This longer view of use potential has recently seen a revival under the sustainability agenda as reported at the 2001 AIA convention.[3] (Plugman, 2001). The research supporting this book also confirms this idea of the general utility of buildings. However, while this encourages adaptation as a serious alternative to demolition and new build, it does not help to determine which new use is best suited to a particular building in a particular location at a particular time. This more complex issue, which has much greater relevance to the problems of today's cities, and is a primary concern of anyone considering the fate of a particular building, is a major part of the material covered by this book.

To investigate the issues relating to which new uses are likely to be functionally and financially viable for an existing building required that as researchers we explore the complexities of existing practice in adaptive reuse. To do this it was necessary to identify and categorise all of the major players involved in making the decisions and implementing adaptive refurbishment projects at present, from financiers to users. It also involved establishing a rigorous framework of criteria for decisions, such as those relating to risk or cost, from which the enquiries could be developed. Some of the detail of these investigative mechanisms is described in the following sections in order to provide the reader with what it is hoped will be a convincing explanation of the results and how they may be used to provide guidance on which use may be most appropriate for a given building. Additionally, this is intended to allow the reader to make better use of the guidance given on project delivery, funding and marketing.

It should, however, be pointed out that the book does not deal with the well-known and established techniques and knowledge bases associated with design, project management, costing and project financial analysis. The literature on these is extensive and of long standing, as are the programmes of study in all these areas. All of these are considered by the author to be as valid and well tested for refurbishment projects as they are for new projects, and as such were not part of the research.

What is provided here is a fresh understanding of how an extensive range of physical and locational characteristics can be considered in a systematic way to provide guidance on what uses are best suited to an existing redundant building and how those aspects of funding, design and project management which are unique to adaptive reuse can be more effectively handled to improve the chances of success in this field. The uses which emerge are much more specific than those normally referred to in planning regulations and are much greater in number. The management issues identified alongside the issues of use choice focus on the problems that arise due to the differing perceptions and interests of those involved in this work. The benefits of assembling project teams from those experienced in refurbishment are also described and discussed.

The book begins by dealing with the topic in the context of the UK; however, a number of references are made to US and Canadian practice which show many consistencies with the UK and the research results developed here.

1.2 Refurbishment practice in the United Kingdom

Scale of the refurbishment market

The scale of the refurbishment market in the UK has been growing steadily since the 1970s. By the mid-1990s refurbishment activity represented 42% of the total construction output (categorised by Government as repair and maintenance), worth in 1995 £21,087,000,000 (National Statistics, HM Government, London) (Figure 1.1).[4] This represents a two-fold increase in this sector since the 1970s, when it represented 22.5% of the total construction output within the UK. The two main categories of this work are housing (56%) and commercial (44%). Within these figures lie most of the refurbishment activity that involves change of use. Government planning application statistics put this at 9% of all building activity, but this may be an underestimate because often large-scale refurbishments must be designated as new building activity, as are all cases where housing is converted to other uses. Thus it is likely that change of use refurbishment is an activity involving expenditure of at least £5 billion annually in the United Kingdom.

Each year some 1.5% of the building stock of the UK is demolished,[5] (Department of the Environment, 1987), mainly to be replaced by new buildings. A further 2.5% is subject to major refurbishment and renovation. Therefore, in any one year no more than about 4% of the national building stock will be in the process of physical change, the rest being

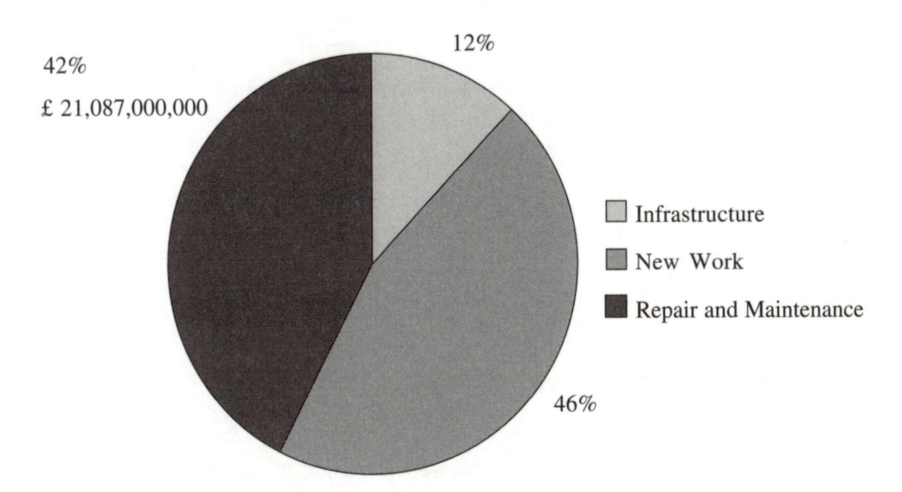

Figure 1.1 1995 construction output by type of work as a percentage of total output of £49,826 million.

Source: Department of Environment 1995 Construction Output Statistics.

subject only to routine maintenance and minor modification. As a consequence, there is a considerable inertia in the available stock of buildings, with a minimum 2–5 year time lag in the adjustments to the supply of new and adapted property to meet changing demands. Changes in the quantity and quality of demand for buildings over the 5–10 year medium term mainly have to be accommodated by the existing stock rather than by new-build developments. Unless building life expectancies reduce dramatically, and replacement rates increase accordingly, the changing requirements of building users must continue to be met by moving to more suitable premises, or through the adaptation and better management of the existing stock.

Change of use characteristics

A sample of planning applications from the London boroughs most active in change of use activity from January 1993 to November 1994 (Barnet, Croydon, Camden, Hackney, Islington, Tower Hamlets and Westminster) shows the broad direction for these 'change of use' proposals. Figure 1.2 shows an origin–destination chart with the planning figures incorporated

		Destination Uses					
		Residential	Retail	Industrial	Office	Other	
Original Uses	Residential	3.5%	1.2%	0.0%	3.2%	0.6%	8.5%
	Retail	1.1%	1.2%	0.3%	2.6%	0.6%	5.8%
	Industrial	7.6%	0.5%	1.5%	3.5%	2.3%	15.4%
	Office	**33.7%**	4.7%	0.2%	1.2%	9.6%	49.4%
	Other	10.8%	0.8%	0.3%	5.5%	3.5%	20.9%
		56.7%	8.4%	2.3%	16.0%	16.6%	100%

Figure 1.2 Origin and destination uses of planning applications involving 'change of use'.

Source: APR database for January 1993–November 1994 for the London boroughs of Barnet, Croydon, Camden, Hackney, Islington, Tower Hamlets and Westminster (n = 344).

under the relevant use class. These are typically broken down into five broad categories: Retail, Office, Industrial and Warehouse, Residential, and Other (Institutional and Leisure). Changes out of office use accounted for 49.4% of all origins, and changes to residential accounted for 56.7% of all destinations. Four types of 'change of use' were dominant: Office to Residential (33.7% of all cases), 'Other' (public buildings, education, hospital) to Residential (10.8%), Office to 'Other' (9.6%), and Industrial to Residential (7.6%), all as shown in Figure 1.2.

This origin–destination approach to 'change of use' analysis, making use of existing statistical data, could provide a valuable but simple management tool for planning authorities in the future. However, for planners as much as for developers, at present 'change of use' refurbishments are tackled in an *ad hoc* manner. Options are considered on a 'project by project' basis with experience remaining private to the individual firms involved. In these circumstances, the development of 'best practice' procedures is limited, with few opportunities to establish guidance for the avoidance of project failures. The knowledge base to inform refurbishment decisions has been developing rapidly but remains inadequate in the context of 'change of use' refurbishment, where there is a critical gap in research understanding and application. Literature surveys show that while there is planning guidance concerning change of use[6] (Department of the Environment, 1991) and the reuse of redundant stock[7] (URBED, 1987), this does not address technical problems and issues. There are no operational methods for the identification of the strategic options for refurbishment to new uses; neither are there established techniques for testing and comparing the relative value of options. As a result, the ability of property owners, user organisations and the property professions to assess the risks and benefits of alternative refurbishment measures remains limited.

Legal framework

However, the current approach is influenced less by management approaches such as referred to above and rather more by legal frameworks. Law regulates the use that is made of buildings. So the history of the adaptation of buildings is, to some extent, a function of these legal regulations, the ways in which they have been interpreted in the past, and the ways in which they are being applied today. Legal controls are slow to respond to changing circumstances. Planning and building regulations tend to operate in a form that, while suited to the circumstances of the past, is often inappropriate to meet contemporary conditions. The regulatory framework may therefore retard the dynamics of change and actually contribute to the problems that it was established to control.

Historically, the laws that control building use have rested on an assumption that the classes of land use and building use types are relatively permanent features, around which an appropriate and beneficial regulatory system for the built environment can be achieved. The regulations of use have been directed negatively, to avoid developments having detrimental effects on public health, welfare, amenity, and existing employment; and positively to control development, to promote economic activity and the creation of jobs, to conduct public consultation regarding proposed developments, to help achieve safety at work, to conserve heritage buildings, and recently to encourage mixed developments where appropriate.

The practical legislative position in the UK is complex, with more than 150 statutory measures for regulating the built environment. However, four major areas of legislation are particularly relevant to the adaptation of buildings. These are:

- Town and Country Planning Act 1990
- Planning (Listed Buildings and Conservation Areas) Act 1990
- Building Act 1984 and Fire Precautions Act 1971
- Health and Safety at Work Act 1974 and Environmental Protection Act 1990.

Each of these Acts has a set of 'Regulations' and 'Orders' which focus on specific areas of concern, targeting particular issues for implementation. From a legal viewpoint, adaptations for change of use are of two basic types. First, there are changes from one use class to another, for example from commercial to residential use. Second, there are changes of use within the same use class, for example the division of a commercial development into a number of permanently separate office units.

Evidently some legislation, particularly those parts that now recognise mixed development and multiple use, has moved towards an understanding of the importance of adaptive reuse. However, this book is not concerned to outline the systems of constraints that apply to this field, but rather to look at how better to effect the changes and understand the opportunities. Accordingly, the next section will attempt to explain some of the basic supply and demand factors working on adapting buildings to new uses.

1.3 Changing use demands and existing supplies

Today the nature and pace of change within organisations and for individuals is having significant impacts on the built environment. Driven by information technology, global competition and environmental concerns, new organisational structures, flexible employment arrangements, novel working practices and changing demands for transport facilities are rapidly emerging. These fundamental developments are resulting in profound adjustments to the demand for, and the use of, urban space. Demand side changes of this kind in the UK, alongside recession, resulted in unprecedented high levels of under-utilisation, long-term vacancy and redundancy in some types of building stock, particularly in the public and private office sectors, by the mid-1990s. Unlettable office space in the City of London alone reached 500,000 square metres by 1995[8] (Gann & Barlow, 1996). In other sectors, such as the rented housing market, there was severe scarcity of supply and further dramatic increases in demand were predicted (DoE household Projections, 1992: 3.5 million additional Households by 2011, 4.4 million by 2016). The achievement of environmentally sustainable urban environments by effective and appropriate change of use of redundant building types to meet the evident new use demands will be a continuing challenge to all involved in the decades ahead. To help in understanding the nature of this challenge, some of the key characteristics of the property market and how these changes are affecting it are discussed below.

The demand for and supply of buildings

Within the property market, the general characteristics of the demand for and the supply of building space are well understood, at both the level of the individual organisation and that of the local property market. From the demand side, organisations have four simple options for procuring the amount, quality and location of space to meet their requirements. They may choose to purchase or rent suitable accommodation from the array of available stock; they may decide to modify, adapt or extend their existing accommodation; they may decide to procure a new or reconditioned building; or they may adopt a mixed strategy that combines elements of each option. Purchasing or renting from the available stock has always been the preferred option, accounting for most of the transactions by far in both the office and residential sectors.

From the supply side, each available building presents an array of different resources to a potential occupier. There will be differences in the physical resources offered, e.g. in floor area, internal spatial arrangements, services, and in the age, quality, character and condition of the building. There will also be locational differences in the resources available, e.g. in the sufficiency or scarcity of transport services, shopping, amenity and leisure facilities, and in the relative availability of suitable employees, customers or business clients. Buildings differ when viewed as a financial resource, with different rent, purchase and rateable values, implying different costs and returns for the occupier. Finally, there will be differences in the range of uses which any building is able to support, and in their potential for flexibility and change.

It is through the matching of the resources demanded by an organisation with the resources supplied by its building and its location that the suitability of an organisation's premises are usually assessed. This matching process underlies the characteristic cycle of 'supply and demand':

- an evaluation of the use of the existing premises, whether over- or under-utilised
- an assessment of future organisational needs and demands
- actions to adjust the property portfolio and its provisions to meet the anticipated needs
- a re-evaluation of the utilisation of the adjusted provision
- an assessment of the next cycle of supply and demand

and so on. In this way the problems of building supply and demand are never permanently resolved, but periodically adjusted over time.

Within the property market as a whole, significant shifts in the balance of supply and demand are commonplace within the lifespan of any building. In a period of recession, the demand for workspace is suppressed, rents level, property values fall, and the vacancy rates in the commercial building stock increase. In a boom period, demand for space exceeds the existing supply, stock scarcity inflates rents and property values, vacancy rates fall to below the 5% level at which normal business moves begin to become difficult, and new buildings are started, to complete often just as recession recurs in three to five years' time[9] (McKee, 1996). Over the 40 years to 1989–90, the general balance of the market favoured suppliers in the UK, and long leases with upward-only rent reviews prevailed. In the 1990s, following an exceptional boom in office building construction, the balance began to tilt towards buyers of space. Pervious booms had not had this effect and it became apparent that there might be a structural change in the demand for space relating to the technological and economic factors referred to earlier and discussed in DEGW's Orbit Reports in the mid-1980s[10] (DEGW, 1985).

Demand-side changes

Over the ten years 1985–1995, demand-side changes included the following.

- A continuing reduction in *manufacturing employment* generally with downsizing of 18%,[11] directly and significantly lowering the gross national demand for industrial building space over these ten years.
- The impact of *information technology*, particularly mobile telecommunications and networked PC systems with dramatically increased processing power, fundamentally changing locational constraints and opportunities, imposing less demands on building services than in the past. The pace of IT developments is resulting in profound adjustments to the demand for and use of office, retail, leisure and transport facilities, and in the criteria for residential locations[12] (DEGW, 1985).

- Changing *organisational requirements* for smaller and more flexible units of usable space, shorter leases, with distributed business operations to meet contemporary demands and the convenience of customer, subcontractor and staff.
- Flexible *employment strategies* including flexible working hours, part-time working and job sharing schemes, shorter-term contracts, annual hours schemes, and the growth in outsourcing in both the private and public sectors as famously discussed by Charles Handy in *The Age of Unreason*[13] and other volumes (Handy, 1991).
- Changing *working practices* that lead to less space per employee through space sharing, degrees of home working, electronic filing, and improved space utilisation generally[14] (Becker *et al.*, in 1991).
- Changing *user expectations* concerning human health and well-being in the workplace, amenity, energy efficiency, and particularly the demand for more natural environments.
- Changing *regulatory requirements* for health and safety, and for maintaining buildings in a commissioned state.
- The professional *facility management* of building stock and its support services, achieving greatly improved levels of efficiency in building use and space productivity overall, thereby reducing the effective demand for building space.

Supply-side effects

These changes in the patterns of demand for building use resulted in a number of supply-side problems, particularly in the early and mid-1990s, including the following.

- Significant levels of long-term *vacancy* and *under-utilisation* in the building stock, particularly in the office sector where unprecedented vacancy rates in excess of 15–20% persisted in major world cities such as London and New York.[15] (Weatherall Green & Smith, 1997). Significant vacancy levels of 2.7 million sq. m of freehold space and 3 million sq. m of leasehold space also occurred in the UK public office sector in 1996, following the adoption of Private Finance Initiative (PFI) criteria for public sector property.
- Significant *redundancy* in some highly specialised building types, for example London hospital buildings and telephone exchanges throughout the UK.
- Significant reductions in the *asset value* of building stock and land generally, with reductions as rents levelled or reduced in response to vacancy rates.
- Significant *underprovision* in key sectors, particularly in the rented housing market where there is severe scarcity of supply, with further

dramatic increases in additional households predicted over the next 15 years.
- Increasing levels of premature *obsolescence* in the commercial building stock generally, driven by functional and locational factors rather than by age, physical deterioration and depreciation.
- Significant degrees of *physical constraint* over use and performance, imposed by the spatial form, servicing regimes and specification of much of the 1970s and 1980s building stock, most severely in the case of shallow plan, fully conditioned office environments in the private sector, and poorly designed and cheaply specified buildings in the public sector.
- Significant dereliction in inner city areas, particularly those areas which have been affected by de-industrialisation, contributing to increasingly *unsustainable* urban environments.

The impact of these changes in both demand and supply on the use to which buildings are put is considerable. The sustainability and even the survival of cities will depend on how successful we are in adapting buildings to new uses. Adaptive reuse is a complex process which requires that the participants in the process have a clear understanding of how to determine what future uses will be most appropriate for a particular building in a particular location and for a given period in time. Section 1.4 provides some insight into the starting points for this process.

1.4 Starting points for change of use*

Basic options

The origins of the process of any single case of adaptation lie simply in the existence of a building that is no longer fully needed for the functions it originally performed or was intended to perform. There are a number of circumstances that can prevail to create this condition, and before considering adaptive reuse for a particular building it is useful to explore which of these circumstances may exist. Some commentary on the options worthy of particular consideration follows.

The rapidly changing pattern of requirements for buildings, and the resulting imbalances in supply and demand, can only be modified in one of two ways: through the adaptive reuse of vacant and under-utilised buildings, or through the replacement of an increasing proportion of

* I am indebted to Professor Bev Nutt for allowing me to modify some of his original text for this section.

redundant stock, often before it reaches its normal financial and physical life expectancy. Where a building has been vacant or under-utilised for a considerable period of time, six basic options are available.

- *Market*: undertaking no further physical improvements, refurbishments or adaptations, and committing no additional investment in the property, but introducing or intensifying arrangements and inducements to encourage potential occupiers to purchase or rent the building.
- *Leave vacant*: deciding to mothball the building in a vacant state until market opportunities improve, perhaps stripping out and maintaining the building shell only, in order to reduce local tax liabilities and to prepare the property for its rehabilitation in due course.
- *Refurbish*: renewing and upgrading the building under its current use class to contemporary standards appropriate to its location, sector and market, so that the property becomes attractive to potential occupiers, improving its marketability for sale or rent.
- *Modify use*: refurbishing and adapting the building to accommodate changing requirements for use and different types of occupancy, within the same dominant use class, perhaps adding ancillary uses.
- *Change class of use*: adapting and refurbishing the building for a new single class of use or to mixed uses, for example from industrial use to mixed small retail, residential and professional office use.
- *Demolish*: redeveloping or selling the site.

These options are explored in Figure 1.3, a diagram developed by my colleague Professor Bev Nutt which explores the potential states of a given item of building stock and shows the three future state possibilities which are the main concern of this book.

Clearly there are **three** dominant possibilities: **adaptations for same use plus ancillary uses, adaptations for mixed classes of use** and **adaptations to a totally new class of use.** Of these three, it was evident from the research that the most serious challenges arose in the last two in relation to the major issues of finance, marketing, design and local authority approval. However, the provision of ancillary uses can raise the same questions of choice and implementation as the more challenging projects and is therefore equally relevant to this coverage.

Having looked at the broad spectrum of possible options for an item of building stock and narrowed this down to those that are clearly adaptive reuse, the next section will look at the participants in the process. This time the objective is not to consider the elimination of some, since whatever the project type, the behaviour and contribution of all participants must be understood as each affects the other and few projects are the result of individual effort by one participant (outside the realm of DIY).

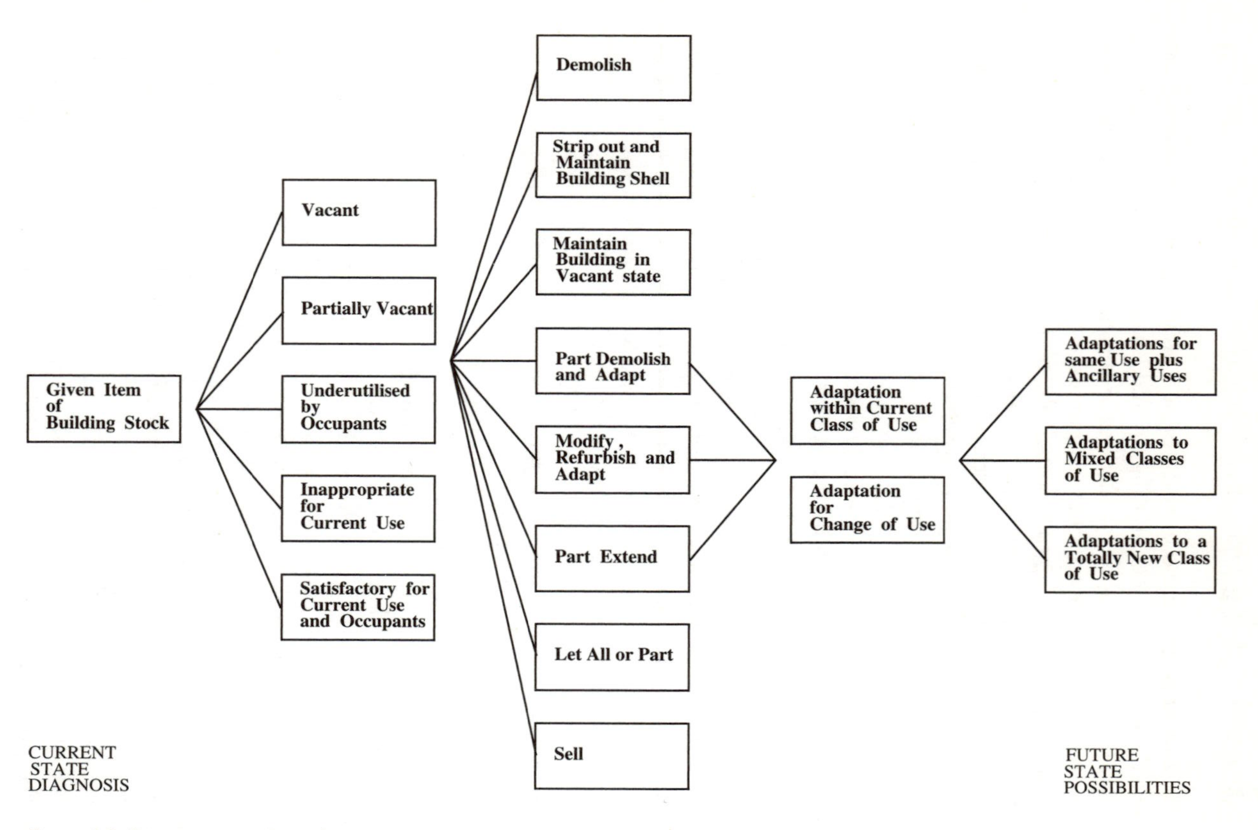

Figure 1.3 Basic options for adaptation.

1.5 Participants in the process

Five key areas of decision

To provide clarity to the research, it proved useful to aggregate all the many individuals and organisations involved in decision-making for the adaptive reuse of buildings into five groups with generically distinctive roles. These aggregations derived from considering the need in any project to:

- secure financial resources
- retain individuals and organisations with design and construction skills
- market the adapted building (owner occupiers may question this, but building market value is always an element of decision-making even within an organisation)
- obtain approvals to the change of use and the detailed design
- satisfy the current and future needs of occupiers.

The participants who carry out these roles are, for simplicity, described as **investors, producers (designers and constructors), marketeers, regulators** and **users.** As these five are involved in all of the major decisions of any project they are clearly, in management terminology, **decision agents.**

Table 1.1 describes in more detail the participants who make up these five decision agents as well as identifying a sixth decision agent, the

Table 1.1 Decision agents for refurbishment for change of use.

Decision agents	*Description and professional affiliations*
Investors	Pension funds, insurance companies, banks, independent investors, professionals who find capital to fund and potentially purchase a building, possibly from IA, ICA, and IRRV members
Producers	Architects, consulting engineers, surveyors, contractors, specialist suppliers, professionals who develop the specification, cost the specification and implement the changes to a building, RICS, RIBA, CIOB and ACE members
Marketeers	Surveyors, agents, professionals who find users for buildings, RICS members
Regulators	Local authorities, English Heritage, DETR, professionals who review the statutory requirements for changes to a building, RTPI members
Users	
Corporate	Large institutional owners and users of buildings
Residential	Individual users of buildings
Developers	Organisations that seek to combine investment, production and marketing in whole or in part, professionals who derive from any of the above-listed professions and others

Source: Bartlett Research.

developer, who combines the roles of investors, producers and marketeers or at least two of these. Developers thereby dominate decisions for projects which they conceive, often reaping efficiency rewards from this near-monopoly but also suffering the consequences of going unchallenged when ill-informed or mistaken.

The decision process in adaptive reuse

Table 1.1 makes it clear that there are a very large number of participants in decision-making in this field. Each participant in turn represents a wide range of interests and perceptions and as such is likely to adjust and change their position with the passage of time. This may not be so remarkable for building projects generally, including new buildings, but the unfolding of events and the discovery of physical problems and opportunities related to the development of an adaptive reuse project is probably more prone to change than for new works. Thus, even major decisions such as choosing the ultimate use for a refurbished building may change as opportunities in the marketplace change or the building is found to be physically different to what had been assumed at the outset. Accordingly, adaptive reuse projects almost never follow a simple logical sequence of decisions from acquisition through design and construction to marketing, as might, sometimes at least, occur in new-build work. Thus the whole decision process is iterative and uneven to some degree, even if the major decision stages can be identified as suggested above in Figure 1.3.

This makes it particularly important to understand the basis of decisions taken in adaptive reuse but also to consider the impact of differing criteria for decisions that may be used by each of the five decision agents. Evidence from the research suggests that the differing focuses and perceptions of these five can cause problems peculiar to this area of activity. Participants in one group can fail to recognise the importance and even the nature of decisions to be taken by other groups. Thus project completions were found to have been delayed significantly in some cases by designer insistence on achieving particular standards of equipment or finish, only to find that the market for the product had collapsed by the time the building was ready to be offered. The failure of investors to influence events differently in such cases or to appreciate that designers were not tuned into market behaviour have in the past led to adaptive reuse bankruptcies.

Further complicating things in this field, it is evident that any one of the decision agents may participate at any stage in the process. Thus regulators such as English Heritage may decide to 'list the building', dramatically limiting its use potentials and often its market value[16] (Scanlon *et al.*, 1994), at almost any stage in the proceedings. More usually this decision will be taken by any of the other decision agents in response to the market and in relation to the location of the building and its physical characteristics.

Regulators will then have the final say on this key decision. Thus it is evident that the increase in demand for residential accommodation in Central London and the decline in demand for office space was, at the time of the research in the mid-1990s, dominating the alternatives in change of use in London, as shown above in Figure 1.2. Perhaps less evident was the regulatory resistance to this direction as structure plans were upset and investor resistance emerged due to loss of value in moving from office use to housing.

These differing perceptions and the variation between the roles and agendas of the decision agents were explored by considering the criteria each must use to assess the viability of the project. This is discussed in section 1.6.

1.6 The criteria for viability

Four key criteria

All involved in adaptive reuse necessarily play some part in making the decisions that determine the scale, the content and the timing of projects. Such decisions have one thing in common: they are ultimately concerned with the viability of a project. This is not to ignore the obvious point that project viability, though central to the longer-term legitimate interests of all major participants, may not be the primary driver for the direction of many decisions. In this regard there can be no doubt that many decisions affecting the viability of a project are taken for reasons of convenience, self-interest or expediency. While not ignoring the fact that such criteria can affect project viability, it was considered that such decision criteria do little to inform us about how to identify the basis for key project decisions but rather are at root a matter of management control and leadership. The latter, while an essential part of effective project management, are not unique to adaptive reuse, nor are such criteria amenable to consistent analysis as they are circumstantial and open to endless variation.

Accordingly, using project viability as the key to identifying decision criteria led originally to the identification of six viability criteria: cost, benefit, value, risk, utility and robustness. However, on further reflection it was seen that because benefit and utility can be expressed in terms of cost or value, the final dimensions of viability could be reduced to a core of four categories:

- the relative COST of options
- the relative VALUE of options
- the relative RISK of options
- the relative ROBUSTNESS of options.

Differing preferences of decision agents

These four viability dimensions provide a basis for describing current decision-making behaviour by the six decision agents in adaptive reuse. This behaviour was examined in the research by analysing decision agent responses to a set of questions in relation to the four dimensions of viability. Figure 1.4 summarises this analysis by displaying the percentage of positive responses to the questions within each viability dimension for each decision agent. Differences such as the 81% positive response of marketeers to value criteria in contrast to the 28% positive response by producers (designers and contractors) are very clear and almost alarming. When one considers that both designers and contractors make daily decisions on both conceptual and detailed matters that affect the market value of the completed refurbishment, this apparent lack of interest in value alongside a dominant focus on cost surely puts project viability at risk. When this characteristic is compared with the producer group's relative disinterest in risk, much lower at 35% than that of any other decision agents, then project managers need to take notice. Investors should also note the lack of interest in robustness of schemes, except by marketeers, which can jeopardise the long-term outlook of their organisations. This kind of information is a clear warning of latent problems within project coalitions and points to the need to guard against isolated decision-making by members of project teams.

	Investor	Producer	Marketer	Developer	Regulator	User Corporate	Viability Totals
Cost	52%	**54%**	49%	42%	NA	**84%**	56%
Value	44%	28%	**81%**	55%	**79%**	65%	**59%**
Risk	**57%**	35%	60%	**61%**	62%	51%	54%
Robustness	35%	39%	71%	48%	49%	55%	50%
Participant Totals	47%	39%	**65%**	52%	63%	64%	

Figure 1.4 Participant viability emphasis: percentages of positive responses to questions.
Source: Bartlett Research.

Table 1.2 Viability emphasis by decision agent.

Decision Agent	*Viability Measure*			
Producer/ Corporate User	Cost			
Marketer/ Regulator	Value			
Investor/ Developer	Risk			
None	Robustness			

Source: Bartlett Research.

A close look at viability criteria and particularly the tendency of different groups to show inconsistent preferences can be critical to understanding where action can most usefully be taken to improve the quality of decisions in adaptive reuse. This is probably the key to understanding the extent and nature of the barriers to communication and to developing the remedies that will improve understanding and open the channels of communication. Table 1.2, which is a derivative of Figure 1.4, summarises the viability measures preferred by sets of decision agents and is a further illustration of the differences in perspective that arise in current practice. The implications of these findings are discussed further in Chapter 3, which expands on the management issues involved in adaptive reuse.

For the participants in adaptive reuse, there are then a number of different perspectives on what issues are most important. Even within this variety, however, it is possible to identify issues which are of overriding concern to all working in this field. These are discussed and summarised in section 1.7 to complete this introduction.

1.7 Key issues of adaptive reuse

Adaptive reuse as a mainstream activity

In the 1990s in London and many other major Western cities, it became evident that significant changes in the economic environment were creating particularly marked imbalances between supply and demand for buildings in various use categories. These economic changes were driven by technological, global trading and management system changes which, further fuelled by recession in the UK in the early 1990s, created a surplus of secondary office space and certain types of industrial space alongside a significant demand for housing and retail space. Environmental concerns grew during this period and the attractions of converting existing surplus space to new uses became increasingly evident.

At the current time, though the office sector is more settled, post-industrial dereliction persists and the attractions of adaptive reuse have become established and embodied into the sustainability agenda. This was firmly endorsed by the work of the Urban Task Force which reported to the UK Government on these issues[17] (Department of the Environment, Transport and the Regions, 1998). This means that this activity is moving towards the mainstream of the built environment, whereas in the past it was dominated by those developers and contractors that might be categorised as risk acceptors, who suffered frequent failures either at planning approval stage or in the marketplace. As it does so, all of the decision agents are likely to seek to control risk more reliably by identifying the key risk issues and developing information and operational management systems to reduce these.

Another factor moving adaptive reuse to the mainstream is seen in relation to the rapid rate of change now evident in all developed economies. This is evident when we look at one of the major impacts of information technology, particularly when aligned with global marketplaces, i.e. the reduction of planning time horizons in businesses and for products. Rastogi[18] (2001) tells us that the future needs of organisations and user groups, in both the private and public sectors, can no longer be forecast with confidence beyond three to five years as product life cycles reduce and working patterns and markets shift dramatically. He relates this phenomenon to the geometric growth in human knowledge and its dissemination. The effect of this on the design and use of buildings which are typically built for lifespans of from 40 to 80 years[19] (Cowan, 1963), depending on function, is profound. In the past it was assumed that an understanding of the intended use of a building provided the appropriate starting point for responsible design. For the past 30 years, this idea has formed the basis for the demand-led architectural brief and its analysis of client requirements as described in the RIBA Plan of Work. Most designers have of course incorporated contingency measures within this process, to help face the indeterminacies of the future. However, the expectation of change to the use for a building, within a strategic approach to design, has been rare[20] (Nutt, 1993).

Accordingly, today it is no longer reasonable to assume that most new-build stock will remain within its original class of use, or that the class of use defines the physical need, throughout its effective physical life. It is also optimistic to assume that most 'change of use' refurbishments, once made, will not be subject to further changes in due course. The frequent adaptive reuse of existing built space may soon become the norm rather than the exception, requiring planning and design procedures that enable built space to be adjusted and re-adjusted to satisfy the rapidly changing patterns of demand. The extent and rate at which the existing building stock is capable of adaptation to support changing uses and requirements will need to be increased significantly to support the challenge of accelerating economic change.

Changing property strategies

Changes in the balance of demand and supply have also led to adjustments in the relative importance of the different strategies within property planning overall. Priorities have tended to move away from the problems of new building procurement towards property consolidation, building use strategies, better facilities management and the reuse and redesign of the existing building stock. The emphasis has shifted from the problems of acquisition to the problems of disposal for many industries. Good examples of this are to be found in telecommunications, where the traditional telephone exchange building has become largely redundant (British Telecom had over 80 million square feet of surplus in the 1990s) and famously the retail bank branches in the UK, the extensive closures of which have led to much public anger – and many new pizza parlours.

In these circumstances, before the commitment of resources to refurbishment is confirmed, it would seem essential to ensure as far as possible that the intended use of the building is likely to remain secure for at least the period necessary to recover the financial capital to be expended on the work. Beyond this, the shrewd investor would be well advised to consider the robustness of his or her decision by looking to the possible alternative uses that might be accommodated at an affordable cost should the initial use choice prove unsustainable. The class of future use tends to be the most critical single factor affecting the development of the brief, the overall design concept, the specification of materials, the resulting costs of adaptation and the expectations concerning financial returns. A use-viability check needs to be undertaken to assess the relative risk of the designated use for the project as a whole. In cases where no change of use is planned, this will serve to reassure investors of the continuing viability of the current class of use for the building and its location. In cases where a change of use is planned, the check will help to confirm that the preferred future use class is superior to other alternative uses.

In summary, owners and managers that are considering the refurbishment of an existing building will face six key questions.

1. What is the use potential and financial value of the building under its present class of use, given current and emerging market conditions?
2. In the current circumstances, is refurbishment within the existing class of use sensible and secure, or should the possibilities for adaptive reuse be considered?

3. If the building is vacant, significantly under-utilised or inappropriate for its current use, what is the property's basic capacity to accommodate change, particularly its 'adaptability potential'?
4. How can the range of potentially viable options for change of use adaptations be identified?
5. What set of characteristics make the building 'more' or 'less' adaptable, and how should its 'adaptability potential' be assessed?
6. How should the strategic and technical viability of proposed options for adaptation to new uses be examined practically, and what decision support systems can be used to assist in the evaluation?

Applying the findings of the research to future decisions

In this section the author has attempted to outline the importance of adaptive reuse and how current practice in this field can be viewed, based largely on UK experience and research. Chapter 2 explores in some detail the question that lies behind the six issues identified at the end of the previous section: which new uses are best suited at a particular time to an available building? The response to this question is explored in terms of physical and locational characteristics by examining a set of near matches between supply and demand using a computer-based comparator. The comparisons are based on the material gathered by looking at the decisions made by decision agents and the criteria used for their decisions.

Chapter 2

Finding viable uses for redundant buildings

2.1 Characterising the available supply

Supply, demand, performance and decisions

The viability of any proposed 'change of use' adaptation, either for a redundant building or for a class of under-utilised property as a whole, needs to be examined against three critical sets of criteria (functional and use viability, technical and physical viability, and economic and financial viability), as outlined in Chapter 1. This chapter sets out the basic decision framework through which the possibilities for change of use can be explored, and the viability of identified options is examined. This search for viable uses for redundant buildings covers the following.

- The **supply characteristics**: the set of physical opportunities and constraints of the building, its location, site, facilities and support services.
- The **demand characteristics**: the set of use requirements by function and specific type of use, describing the demand-led needs of user and organisation.
- The **performance requirements**: the interface between supply and demand, matching the set of physical provisions with the set of operational requirements.
- The **decision procedures**: the means by which the use viability, physical viability and financial viability of alternative options for change may be assessed.

Each of these four aspects will be examined in turn in this chapter. First, the supply characteristics that need to be considered during any refurbishment project are described and profiled. Then the generic demand characteristics of potential use are scrutinised and classified, within the specific context of change of use refurbishment. Following this, the computer-based Use Comparator system is described and an example is given. Through this system, the supply characteristics of a redundant building and the demand characteristics of a potential use and the related performance requirements can be matched. Finally, a framework to assist in the process of decision-making is proposed alongside a general consideration of aspects of financial viability assessment.

Identifying generic characteristics

When a building appears to be redundant under its current class of use, the first priority is to become aware of the circumstances that may make it more or less suitable as a candidate for adaptation to support new uses. The identification of the generic characteristics that can impact on the adaptation of a building is the essential first step in searching for alternative and viable uses for redundant property. In this search it should be expected that many property characteristics will be largely neutral, with little measurable effect one way or the other on project viability. However, it is likely that some factors will have negative effect, tending to increase project risk and limit refurbishment possibilities overall. Many factors can have a positive influence, in that they promise to increase or enhance the opportunities for adaptation. So, in characterising the available supply of candidate buildings for adaptation, it is essential to:

- identify all *negative* factors for modification or elimination
- target the *positive* characteristics for promotion and exploitation
- recognise those issues that have largely *neutral* effect.

The general viability of adapting a building to support new uses will depend on the degree to which the negative characteristics and constraints might be overcome on the one hand, and the positive characteristics and opportunities might be enhanced and exploited on the other hand. The evaluation of the constraints and opportunities that are afforded by the 'as found' characteristics of a redundant building is the basic starting point in any consideration of adaptive reuse.

Studies conducted in the mid-1990s[21] (Gann & Barlow, 1996) indicate that the relative ease with which redundant offices may be converted into flats is dependent on seven main characteristics of a building:

- the size, height and depth of a building
- the type of building structure
- the building's envelope and cladding
- its internal space, layout and access
- the building's services

- the provision for acoustic separation
- fire safety measures and the means of escape.

At a more detailed level, the general characteristics of building stock that can, in principle, influence refurbishment, conversion and 'change of use' decisions are summarised in Table 2.1, compiled from known published sources[22] (Sigworth Wilkinson, 1967). The table lists the main physical factors relating to the structural, constructional, spatial, environmental and

Table 2.1 General building stock characteristics.

Characteristic	*Factors*
Structural	Type and condition of structure Floor load capacity Structural grid and section dimensions
Constructional	Type of construction and materials Cladding system and fenestration Partitioning and finish Fabric age, condition and maintenance
Spatial	Spatial configuration Floor plate size and floor depth Core and riser size and locations Entrance and floor access arrangements Fire escape provisions Planning grid dimensions
Environmental	Type of environmental regime Orientation and energy profile Lighting and ventilation arrangements Environmental control systems
Servicing	HVAC system and distribution Duct space capacity Plant capacity and controls Power load capacity IT arrangements and capacity Plumbing system arrangements Services age and condition
Financial	Market constraints and opportunities Property value and land value Exchange value and tradability Rental value and lettability Rate of return and costs-in-use profile Maintenance cost profile
Operational	Locational and site characteristics Transport, access and parking Tenure arrangements Security, health & safety arrangements Usability, flexibility, manageability

servicing characteristics that need to be examined for their possible positive or negative affect on 'change of use' viability. It also includes the main financial and operational issues that need to be taken into account.

Elements changed during refurbishment

While the importance of specific characteristics will vary widely project by project, there is a need to clarify which of the general characteristics listed above are normally significant during the adaptive reuse of buildings. The case studies, field investigations, questionnaire surveys and structured interviews that were undertaken during the Refurbishment for Change of Use Project (as summarised in the Appendix) indicate the relative importance of the characteristics, as perceived by those involved, and their impact on 'change of use' decisions. These investigations have identified the characteristics that need to be changed most frequently in conversions to new uses. They have also distinguished which building characteristics are preferred by the different types of decision agents involved in refurbishment.

The physical aspects of buildings that were most commonly changed during refurbishment are shown in Table 2.2, broken down by class of use. It can be seen that the most frequently changed building elements were air conditioning and the heating and ventilation services. All building services were usually replaced during refurbishments for 'change of use'. The initial 'as found' condition of services and plant therefore had little direct effect, either positive or negative, on project viability. The second most frequent set of physical changes related to the means of fire escape, across all types

Table 2.2 Building elements changed during refurbishment.

Type of element	*Elements changed during refurbishment (as % of all responses, n = 127)*			
	Retail (%)	*Office (%)*	*Industrial (%)*	*Residential (%)*
Building foundation	2	6	13	7
Superstructure	14	24	14	23
Cladding material	26	36	20	20
Window/wall ratio	17	23	7	39
Roof type	7	17	17	23
Adjoining structure	13	11	6	22
Means of escape	39	**55**	27	**52**
Heating	35	**58**	31	**58**
Ventilation	39	**56**	29	37
Air conditioning	34	**57**	12	10
Building access	34	29	26	35
Core layout	16	48	13	38

of use. This finding supports the results of earlier work which indicated that meeting the fire regulations was the most important consideration, particularly in the conversion of offices to flats. A surprising finding, substantiated by further field investigations, was the extent to which the core layout was changed, particularly during office and residential refurbishments, where 48% and 38% of cases respectively involved physical changes to the building core. This issue is considered in detail in section 2.5. Other frequent changes include building access, which was modified in 26–35% of cases, and cladding, which was replaced or modified in 20–36% of all cases.

Factors affecting value

The preferred characteristics of refurbished buildings in relation to post-refurbishment value are indicated in Table 2.3. The most important physical characteristics affecting the marketability of a refurbished property were building character, period features, floor to ceiling height, and window size. This was a common pattern in all use sectors except industrial. The relative effect of physical building characteristics on the ability to sell within each use class is shown in Table 2.4, building access being the most significant factor. Overall, the building characteristics preferred by developers were very similar to those of the producer and marketing groups, with emphasis on location (61% of all cases), floor plate size (54%), floor to ceiling height (48%), transport access (42%), and particularly building character, period features, window size and local amenities, in the case of office and residential developments. Two locational characteristics were dominant in relation to the saleability of property after conversion: access to public transport and quality of local amenities, as shown in Table 2.5. Users' preferences were remarkably consistent with

Table 2.3 Characteristics affecting market value.

Characteristic	*Positive effect on market value (as % of all responses, n = 176)*			
	Retail (%)	*Office (%)*	*Industrial (%)*	*Residential (%)*
Building character	34	**71**	18	**68**
Period features	25	**58**	6	**62**
Listed building status	6	23	5	42
Brick cladding	12	26	23	34
Curtain wall cladding	2	26	12	3
Stone cladding	6	25	8	15
Floor to ceiling height	23	25	43	26
Size of windows	35	**65**	9	43

the marketing and developer groups, with positive external factors including building character, period features, size of windows, car parking and transport access being of paramount importance. Within the building, floor to ceiling height, floor plate size and the configuration of core areas were seen as the more important features.

Table 2.4 Physical variables affecting sale.

Physical characteristic	*Effect on ability to sell by sector (as % of total responses, n = 67)*			
	Retail (%)	*Office (%)*	*Industrial (%)*	*Residential (%)*
	Developer group			
Cladding material	7	37	30	30
Window/wall ratio	13	37	10	43
Roof type	10	13	20	23
Adjoining structure	23	30	23	40
Means of escape	23	30	20	13
Heating	20	43	17	43
Ventilation	20	43	20	23
Air conditioning	7	47	7	0
Building access	30	30	33	37
Core layout	13	43	13	10
	Marketing group			
Cladding material	17	**57**	40	31
Window/wall ratio	29	40	46	49
Roof type	46	**69**	17	31
Adjoining structure	46	**57**	37	**57**
Means of escape	34	**71**	37	40
Heating	23	**74**	46	**60**
Ventilation	26	**71**	**57**	20
Air conditioning	23	**89**	20	14
Building access	**63**	**74**	**69**	43
Core layout	29	**83**	31	23
	All respondents			
Cladding material	14	**58**	38	34
Window/wall ratio	32	29	15	38
Roof type	22	46	37	40
Adjoining structure	37	**55**	32	**52**
Means of escape	31	**63**	32	31
Heating	23	**63**	35	**54**
Ventilation	25	**74**	43	23
Air conditioning	17	**60**	15	9
Building access	**52**	**71**	**57**	42
Core layout	25	38	26	18

Table 2.5 Locational characteristics.

Aspect	*Locational characeristics affecting supply (as % of total responses, n = 67)*			
	Retail (%)	*Office (%)*	*Industrial (%)*	*Residential (%)*
	Developer group			
Access to public motorway	43	47	37	**60**
Access to motorway	27	**57**	**53**	10
Postal district	7	27	7	**50**
Quality of local amenities	33	43	23	**70**
	Marketing group			
Access to public motorway	**63**	**89**	**51**	**63**
Access to motorway	31	**63**	**80**	**63**
Postal district	29	**57**	14	11
Quality of local amenities	**54**	**69**	29	**66**

Preferred structure and dimensions

The preferred type of structure for 'change of use' refurbishment was firstly steel frame, and secondly load-bearing brickwork, as shown in Table 2.6.

The survey and field study results produced a comprehensive account of the types of structure and fabric, structural grid dimensions, floor to floor heights, floor loading capacities and overall building depths that were preferred by those undertaking refurbishments to retail, office, industrial and residential uses. Summary details of the dimensional characteristics preferred by the developer and producer groups are shown in Table 2.7. It is interesting to note that the developer group, which we know is focused on value, in general prefers significantly greater dimensions both vertically and horizontally to the cost-focused producer group. The market value impacts of steel and brick structures are also notably different between the two groups. In Chapter 3 the project management problems created by this kind of difference in perception are explored further.

Physical profiling

As a result of the questionnaire surveys and field investigations, the inventory of factors shown in Table 2.1 is reduced to a shorter list of key non-service physical variables that should be targeted for detailed examination within the context of a specific project. Firstly, 25 general building physical characteristics that can in principle be modified or enhanced during adaptation are selected, grouped in one of three main categories:

Table 2.6 Preferred type of structure.

Type of structure	*Preferred structural type by reason (as % of responses overall)*					
	Project cost (%)	*Project duration (%)*	*Project disruption (%)*	*Market value (%)*	*Technical choices (%)*	*Other (%)*
	Producer group responses					
Load-bearing stone	1	3	3	15	1	0
Load-bearing brick	30	13	19	18	21	1
Steel frame	34	40	38	18	29	0
Concrete frame	23	11	18	13	16	0
Other	0	0	0	2	1	0
	Developer group responses					
Load-bearing stone	3	3	3	13	3	3
Load-bearing brick	27	23	13	37	7	0
Steel frame	33	43	37	30	23	0
Concrete frame	17	20	13	17	17	0
Other	0	0	0	0	3	0

Table 2.7 Dimensional characteristics.

Refurbishment sector	*Preferred dimensions (average of all responses)*			
	Building depth (m)	*Floor to floor height (m)*	*Structural grid Width (m)*	*Depth (m)*
	Developer group			
Retail	21.7	4.3	6.4	21.7
Office	14.7	3.3	7.0	7.6
Industrial	28.3	6.5	23.3	36.0
Residential	17.8	2.6	5.5	8.3
	Producer group			
Retail	19.3	4.0	7.5	7.5
Office	16.2	3.2	7.0	8.3
Industrial	25.7	5.4	9.0	7.5
Residential	12.0	2.7	5.0	11.0

- location and site (five variables)
- space (12 variables)
- fabric and structure (eight variables).

These groupings, their variables and structure, are illustrated in Figure 2.1. It must be remembered that this is a generic list of key physical characteristics, and may not include all of the relevant characteristics that interact, and need to be taken into account, when assessing the viability of a particular refurbishment project.

Figure 2.1 provides a basic tool for profiling the key supply-side characteristics of a candidate building for 'change of use' adaptation, for both the initial 'as found' state and the possible set of 'future' states after conversion. The first column – existing profile – provides the framework for surveying and examining each of the key physical characteristics in turn, reviewing their potential positive, negative or neutral impact on project viability overall. Judgements are required for each element and group of elements as to whether or not to:

- leave as found and maintain
- modify and upgrade
- fundamentally change and reform.

The second column in Figure 2.1 – possible profiles – can be used in a similar way, providing a framework for the consideration of combinations of physical change possibilities, and the comparison of the alternative profiles that would be achieved.

Having profiled general building characteristics, it is important then to consider the engineering services. These can be similarly profiled by breaking into two parts, mechanical and electrical services, and then into 12 components as follows.

Mechanical

- Heating systems
- Ventilation systems
- Air conditioning
- Water supplies and systems
- Sewage and other drainage
- Fire sprinklers
- Gas supplies

Electrical

- Lighting systems
- Small power supplies

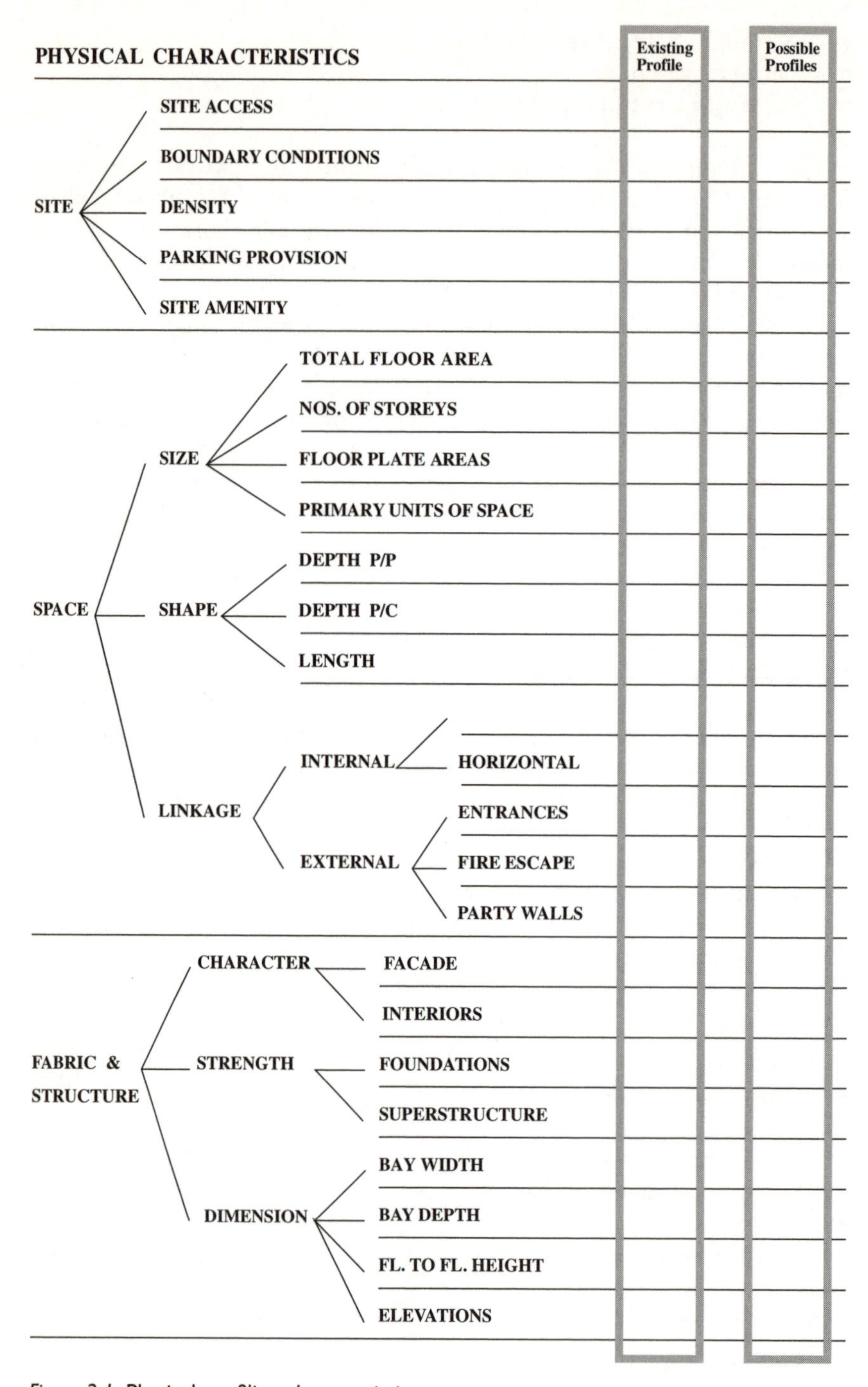

Figure 2.1 Physical profiling characteristics.

- Electrical incomers and mains supply
- Standby power
- Security, control and alarm systems.

It is consistently the case across all use categories that mechanical and electrical systems are the most often changed during refurbishment, as shown in Table 2.2, and it is also the case that these have a dominant effect on sales, as shown in Table 2.4. Again it is important to consider in each case whether these components should be:

- left as found and maintained
- modified and upgraded
- fundamentally changed and reformed.

The examination of the viability of physical change possibilities is described further in section 2.5.

This section has attempted to deal with issues relating to the supply of buildings for refurbishment as informed by the research. In section 2.2 the focus will change to issues of demand by considering possible uses for available buildings.

2.2 Possible uses for available buildings

Demand characteristics of differing uses

In the previous sub-section the key physical and locational characteristics of existing buildings are described and evaluated and many of the research results on user preferences are outlined. The origins of these characteristics are to be found, of course, in the design of the building when new, at which time the use to which the building was to be put was a fundamental factor. These 'demand characteristics' of the building when new are typically based on both a broad generic use for a building, fashions and technology extant when designed and the unique requirements of the original user. Finding new uses for such buildings, which have also usually undergone various modifications through time, requires then an understanding of the demand characteristics of possible new uses as well as a means of comparing these with the characteristics of the building being considered. This section introduces the first of these issues and discusses those characteristics of buildings and locations that can be specifically associated with different uses. The details of these characteristics and the question of comparison with the supply characteristics of any particular building in a particular location are dealt with in section 2.3.

At the outset of a new project, one of the greatest risks to a client is that they create a building that performs badly functionally and financially. An inappropriate brief can lead to an oversized, over-strong, over-decorated, over-serviced and over-expensive building, or the opposite. To avoid this, experienced clients develop clear criteria for everything from floor strengths and ceiling clearances to distances from public transport. Part of the work of the research was to draw on this experience in new building work in order to find out what clients (users) and producers considered to be appropriate criteria for different uses across a wide range of use types. With these values and this information available alongside the findings on preferences for such things as 'building character', as seen in Table 2.3, it became possible to bring together a framework of characteristics relating to specific uses. Locational characteristics were similarly identified and both of these have been built into the comparator as described in section 2.3. Developing this detail is a major part of what is needed to understand better the specifics of the demand side of change of use. However, the other major part remaining is to get a clearer view of what types of use should be considered than is provided by traditional broad planning categories. These may need to be understood for the purposes of meeting planning requirements, but are often far too general to address the real complexities of both the private and public sectors.

Extending the scale of use categories

The case studies in the research showed that the 17 land use categories of the Use Class Order (UCO) system employed by planners are simply too broad to relate to physical characteristics. For instance, UCO A1 – Retail Shops covers everything from a newsagent to a major department store, though they require quite different buildings and locations. Similarly, the CI SfB Building Type categories used by architects relate too much to physical types to relate to actual building uses. In contrast, the over 500 economic activities identified in the Standard Industrial Classification (SIC) system used internationally by economists are in total too detailed for practical use, and many categories have no direct relevancy for buildings. The researchers therefore developed and tested an alternative use classification system, specifically tailored to the issues of the adaptive reuse and mixed use of buildings.

The new framework, which is an EU-compatible simplified version of the SIC system developed specifically to permit comparisons with physical and locational characteristics, employs 77 distinct economic uses that may be directly related to the UCO and CI SfB systems.

The 77 uses were found by asking four questions of each of the over 500 SIC identified economic activities. These were as follows.

1. Does the activity relate to one specific dedicated building space or building? (Clearly deep-sea fishing does not relate to any one building type.)

2. Are types of activity known to be relevant to characteristics for existing general categories such as office? (Finance offices are different to general offices for reasons of location alone.)
3. Is size important, such as in the case of small retail shop or department store?
4. Are specific and different physical and locational characteristics identifiable for this activity?

This Use Class Framework and its relation to the UCO and CI SfB systems are detailed in the Appendix. With this framework it becomes possible to consider comparing the physical and locational characteristics, or **supply**, of available buildings to the building characteristics needed, or **demand**, for a particular use. In fact, the characteristic profiles for each of the 77 uses has been determined and is the basis for the comparisons to available supply that are discussed subsequently.

Relating uses to available buildings

With 77 uses identified in the new classification system, each of which has a preferred set of physical and locational characteristics, it becomes possible, if complex, to compare a given building and its characteristics to the characteristics of each of these 77 basic uses. The number of comparisons required, even if the characteristics are reduced by identifying which are the most critical to compare, suggests that this should be done using computer technology. This has led directly to the development of a decision aid called a Use Comparator, which operates through the use of a standard spreadsheet application programme. The Use Comparator and its capabilities are described in some detail in section 2.3.

2.3 Identifying options with the Use Comparator

Appropriate criteria for supply and demand comparison

Of the four questions which were applied to the fundamental SIC list of economic activities, the fourth question, relating to 'specific and different physical and and locational characteristics' (section 2.2), is the one which provides the most significant link between supply (the existing building) and demand (the 77 economic uses that require a particular building). Thus, to compare an available building to one of the 77 uses it was necessary to identify which of the many characteristics considered by the research were most important in determining use. This was done by considering the results of questionnaire surveys, as summarised in section 2.1, and by identifying from the case studies which characteristics were consistently found to be dominant concerns (whatever the use) or which were fundamental determinants of choice and largely unchangeable without major expenditure.

These considerations led to the identification of 12 key characteristics plus one which is statutory and relates to whether the Use Class Order is Industrial. This latter factor is included in order to eliminate consideration of uses which are too specialist to have any likely generic connection to other buildings. All 13 characteristics are discussed below in more detail, as is the question of how each of these is made use of within the comparator system and how each is measured. The intention of the system is to allow significant measurable characteristics of each of the 77 'uses' to be compared with those same characteristics of any given building. It then shows which uses most closely match the characteristics of the building. These are listed with the closest matches appearing at the top of the list. This allows decision agents to consider economic, financial, social and political factors and set these alongside the comparator results. The system is not judgemental in its use of the data, and is quite unlike the scoring systems for buildings such as Building Quality Assessment, Real Estate Norm or BREEAM inasmuch as it is not used to identify the best building but merely the best fit to usage.

The contents and structure are fully described in the following subsections, and an example of their application to a particular building is fully explained.

Functional framework of the comparator

The Use Comparator is a decision aid that helps the decision-maker to:

- **eliminate** all non-viable change of use options
- **converge** on a set of possible and potentially viable uses
- **select** the principal options for adaptive reuse that warrant detailed appraisal.

It should be noted that detailed appraisal of technical and financial viability within the specific circumstances of any given project is not a part of the comparator function and would be done using established business case methods of the sort that would apply to any investment. The 13 characteristics are arranged such that there are two stages of comparison through which the building proceeds. The first stage is an 'eliminator' which reduces the numbers of uses considered by the second stage, which goes through the full detail of each of the remaining characteristics.

In the **first stage**, unviable options are eliminated for all use class types where:

- the aggregate demand-side requirements for any given use are significantly greater than the supply-side provisions of the item of building stock

- there is a significant mismatch between demand and supply in relation to one or more key factors over which the decision agent has no control or little power to reverse.

In the **second stage**, the potentially viable set of use options are identified where:

- the demand requirements for possible uses closely match the supply provisions of the existing building
- the supply provisions of the building are greater than the demand requirements for use, although these options may prove to be diseconomic.

The comparator system has been applied to and field-tested against a sample of buildings, including seven completed projects, six ongoing refurbishment projects, and many redundant vacant buildings. In all cases the theoretical results were consistent with expert practitioner expectations and in many cases they identified unexpected possibilities. Further refinement and validation of the system will be achieved as usage develops.

Comparator Stage I characteristics

Five characteristics are used in the first stage of the comparator. As the purpose of this stage is to eliminate uses which are significantly inconsistent with the building or location being evaluated, three of these characteristics are measured in a simple binary manner as follows.

Characteristic 1: Use Class Order
Is the location used for Use Class Orders B2 to B7 – Industrial or not? This eliminates a very special category to which planners very seldom allow a change.

Characteristic 2: Hostile factors
Is the location hostile to most people's normal activities by reason of excessive noise, smell, hazard, or mess such as might occur with, say, fish canneries, scrapyards, etc. or is it not? This eliminates a range of important possibilities from housing through to most retail.

Characteristic 3: Tenure
This is a simple question about possible tenure complications, and simply asks if whole or partial tenure is required for the use. This allows the system to distinguish between mixed use and single use situations.

The final two characteristics in Stage 1 deal with two key physical measures which are fundamental in allowing or disallowing certain uses, and for which four distinct values are chosen in each case. These are as follows.

Characteristic 4: Slab to Slab Height
What clearance or slab to slab height does the building provide or the use require? The clearances in each of the four categories are known from the research surveys and reference to specialists to relate to particular uses, and the minimum figures in each category are lower limits of acceptability.

Characteristic 5: Strength
What structural strength does the building provide or the use require? As with clearance, the figures used in each of the four ranges is known to relate to particular uses and the minimums are at lower limits of acceptability.

Tables 2.8 and 2.9 provide greater detail on the relationship between slab to slab height and use and floor strength and use. Many of these values are confirmed by reference to Table 2.7, which summarises the preferences of developers and producers.

The Stage 1 comparator characteristics are summarised in Table 2.10. This shows clearly the simple binary questions associated with the first three characteristics and the quaternary scale which occurs first here but is typical to the Stage 2 characteristics.

Comparator Stage 2 characteristics

Eight characteristics are used in the second stage of the comparator. At this point in the process all of the extreme mismatches of use and building

Table 2.8 Slab to slab height.

Slab to slab height	*Description*
> 4.65 m	Industrial uses, typically B2—B8 and some B1, D2
3.65–4.65 m	Retail uses, typically A1 and A3
2.75–3.65 m	Office uses, typically B1 and A2
2.3–2.75 m	Residential uses, C3, C1, C2 and D2

Table 2.9 Structural strength – floor.

Strength	*Description*
>10 kN/m	Industrial, warehouse uses, B2—B8 and some B1, D2
5–10 kN/m	Light industrial uses, B1
3–5 kN/m	Retail/office, hospital uses, B1 A1, A2, A3, C2 and D2
<3 kN/m	Residential and other uses, C3, C1, C2, D1 and D2

Table 2.10 Stage 1 comparator characteristics.

UCO	*Hostile*	*Tenure*	*Height (m)*	*Strength (kN/m)*
B2–B7	Hostile	Whole	>4.65	>10
Other	not Hostile	Partial	3.65–4.65	5–10
			2.75–3.65	3–5
			2.3–2.75	<3

have been eliminated and the system is now aimed at finding close matches between supply and demand. To achieve this in relation to both other physical characteristics and locational characteristics, the comparator considers four characteristics in each case. In all eight of these a four-part scale has been used to give some range to the measurements, which are often essentially subjective, and to avoid the temptation to use a neutral point when scoring a building. This helps to ensure that all of the detail provides a direction to the final choice.

Physical characteristics

Characteristic 6: Fabric specification quality

This characteristic deals with both the **quality of the exterior and the interior finishes.** The measurement of these is based on using a specification assessment box as illustrated in Table 2.11. The importance of this characteristic was shown clearly in survey results, as indicated in Table 2.3, and was also apparent from the cases studied.

Characteristic 7: Building character

This deals again with both **exterior and interior factors relating to the strength of character expressed in facade and interior components.** The measurement of these is similarly based on the use of an assessment box, illustrated in Table 2.12. Again, Table 2.3 informs the choice of this characteristic alongside other data.

Table 2.11 Specification assessment box.

Specification materials and finishes	Exterior unique quality	Exterior standard quality
Interior unique quality	4	2
Interior standard quality	3	1

Table 2.12 Building character box.

Building character	Strong facade	Weak facade
Strong interior	4	2
Weak interior	3	1

Characteristic 8: Depth of floor plate

This deals with the predominant **perimeter to perimeter depths required for a particular activity.** There are four depth ranges, which are described more fully in Table 2.13. The values used are again based on the research questionnaire responses and expert advice. The depth measurements are as shown in Table 2.13 and are in part informed by the data shown in Table 2.7. No particular depth is preferable overall; it is merely more suitable to a particular use as shown.

Characteristic 9 External and core access

This refers to whether there are **single or multiple accesses from the outside to the building and whether there are multiple or single cores within the building.** Multiples score higher in the measurement system detailed in Tables 2.14 and 2.15 because more uses can be found for buildings with this characteristic even if all accesses are not exploited in relation to a particular use.

A summary of these physical characteristics is shown in Table 2.16.

Table 2.13 Depth of floor plate.

Depth of floor plate	*Description*
>24 m p to p	Industrial, warehouse uses, B2—B8 and some B1, D2
18—24 m p to p	Light industrial uses, B1 and some B2
12—18 m p to p	Retail/office, hospital uses, B1, A1, A2, A3, C2 and D2
≤12 m p to p	Residential and other uses, C3, C1, C2, D1 and D2

Table 2.14 Access.

Access	*Description*
Multiple external and multiple cores	More than two external entrance options and more than two cores
Single external and multiple cores	Single external entrance and more than two cores
Multiple external and single core	More than two external entrance options and single core
Single external and core	Single external entrance and single core

Table 2.15 Access assessment box.

Access	Multiple external	Single external
Multiple cores	4	3
Single core	2	1

Locational characteristics

Characteristic 10: Street Characteristics

Degree of integration with other streets and urban features is the basic measure for this. Quiet residential streets are generally measured as having low integration on this scale; busy commercial streets, a high level of integration. Table 2.17 describes the scale for this characteristic. The concept of integration referred to is derived from the work of Professor Bill Hillier on Space Syntax which he develops further in his most recent book *Space is the Machine*[23] (Hillier, 1998). Some of the material on which this table is based is shown in Table 2.5. Again there is no preferred value but simply differing characteristics for differing uses.

Table 2.16 Physical characteristics.

Specification	*Character*	*Depth*	*Access*
Exterior/interior unique quality	Strong facade and interior	>24 m p to p	Multiple external and cores
Exterior unique quality	Strong facade	24 m p to p	Multiple cores
Interior unique quality	Weak facade	18 m p to p	Multiple exterior
Exterior/interior standard quality	Weak facade and interior	≤12 m p to p	Single external and cores

Table 2.17 Street characteristics.

Street characteristics	*Description*
Highly integrated/city centre	Typical high street or A type road for entire road network
Highly integrated/ borough centre	Typical high street or A type road for local network
Partially integrated	Streets immediately off the high street or main roads within a local network
Poorly integrated	Streets within the local network

Characteristic 11: Amenity assessment

Both social and physical amenities are included in this measurement, which is aimed at **good leisure and retail options at one extreme and dereliction at the other.** Tables 2.18–2.21 elaborate the detail of this and indicate the basis of the measurement. These measurements are based on the views of the decision agents questioned and the evidence available from the six case studies done.

Characteristic 12: Public transport

Ready access to Underground, bus and railways is the essence of this measure. The five-minute walking distance is used as a fundamental metric and this reflects the kind of provision that was found to be at the limit which influences location choices in London. Table 2.22 provides further

Table 2.18 Local amenity assessment box.

Local amenity	Strong social	Weak social
Strong physical	4	3
Weak physical	2	1

Table 2.19 Social amenity assessment box.

Local social amenity	Strong retail	Weak retail
Strong leisure	Strong social 4	Weak social 3
Weak leisure	Weak social 2	Weak social 1

Leisure: eateries, parks, cinemas, clubs, theatres, etc

Table 2.20 Physical amenity assessment box.

Local physical amenity	Low Vacancy (<20%)	High Vacancy (>20%)
Dereliction low (<20%)	Strong physical 4	Weak physical 2
Dereliction high (>20%)	Weak physical 3	Weak physical 1

detail. Table 2.5 summarises some of the data which support the relevance of particular transport provisions for different categories of use.

Characteristic 13: Private transport
This relates to **ready access to the road network, especially motorways and an airport**. The main elements of this are described in Table 2.23 and the sources are again as indicated in Characteristic 12.

Table 2.21 Amenity – local quality.

Amenity – local quality	*Description*
Strong social and physical	Easy access to multiple leisure and retail options with no noticeable vacancy or dereliction in immediate area
Strong social and weak physical	Easy access to multiple leisure and retail options with some noticeable vacancy or dereliction in immediate area
Weak social and strong physical	Some leisure and retail options with no noticeable vacancy or dereliction in immediate area
Weak social and physical	Few leisure and retail options with noticeable vacancy or dereliction in immediate area

Table 2.22 Public transport.

Public transportation	*Description*
3 forms within 5 minutes of site	Bus, Underground and rail
2 forms within 5 minutes of site	Bus and/or Underground and/or rail
1 form within 5 minutes of site	Bus and/or Underground and/or rail
0 forms within 5 minutes of site	No bus, Underground, rail service within 5 minutes

Table 2.23 Private transport.

Private transportation	*Description*
Arterial road and airport – ready access	2 or more A roads or motorways with access to airport
Arterial road – ready access	2 or more A roads or motorways
Arterial road – limited access	2 or more A roads or motorways
Arterial road – remote access	Single A road or motorway

A summary of the locational characteristics is given in Table 2.24.

Use of the comparator – an example

The Gerrard Street Telephone Exchange (TE) is one example of a potential change of use building. While this is an unusual building in some respects, it highlights many of the issues raised in more conventional adaptations. It is an unusual building because it has many fixed features that would need to be worked around, particularly existing switching equipment occupying a few of the floors and the cable chambers in the basement area. These issues can all be taken into account in the Use Comparator.

Table 2.24 Locational characteristics.

Street	*Amenity*	*Public transport*	*Private transport*
Highly integrated/city centre	Strong social and physical	**3** forms within 5 minutes	A road and airport-ready access
Highly integrated/borough centre	Strong social	**2** forms within 5 minutes	A road-ready access
Partially integrated	Strong physical	**1** form within 5 minutes	A road-limited access
Poorly integrated	Weak social and physical	**0** forms within 5 minutes	A road-remote access

The review process takes a few minutes and will be described in brief starting with the five characteristics in the first stage.

First stage characteristics

The purpose of this stage is to identify those features that will eliminate certain potential uses. These characteristics tend to be features that are not changeable in either the short or medium term. The floor plan of the ground floor (Figure 2.2) is used to illustate a few of these characteristics. The elevation (Figure 2.3) and the location map (Figure 2.4) give further indications of the context of the building. The first stage characteristics for Gerrard Street TE are as follows.

1 What is the existing Use Class Order?
   - Gerrard Street TE is considered *sui generis* under the UCO system as it does not fit within any of the defined uses. The UCO is **other UCO.**
2 Is the location or activity hostile to people?
   - Gerrard Street TE has one face onto the pedestrianised Gerrard Street in the centre of London's 'Chinatown' and another (the opposite side of the building) on quieter but commercial Lisle Street. There is traffic along Lisle Street, but it is typical for a site in this location. The activity rating is **not hostile.**

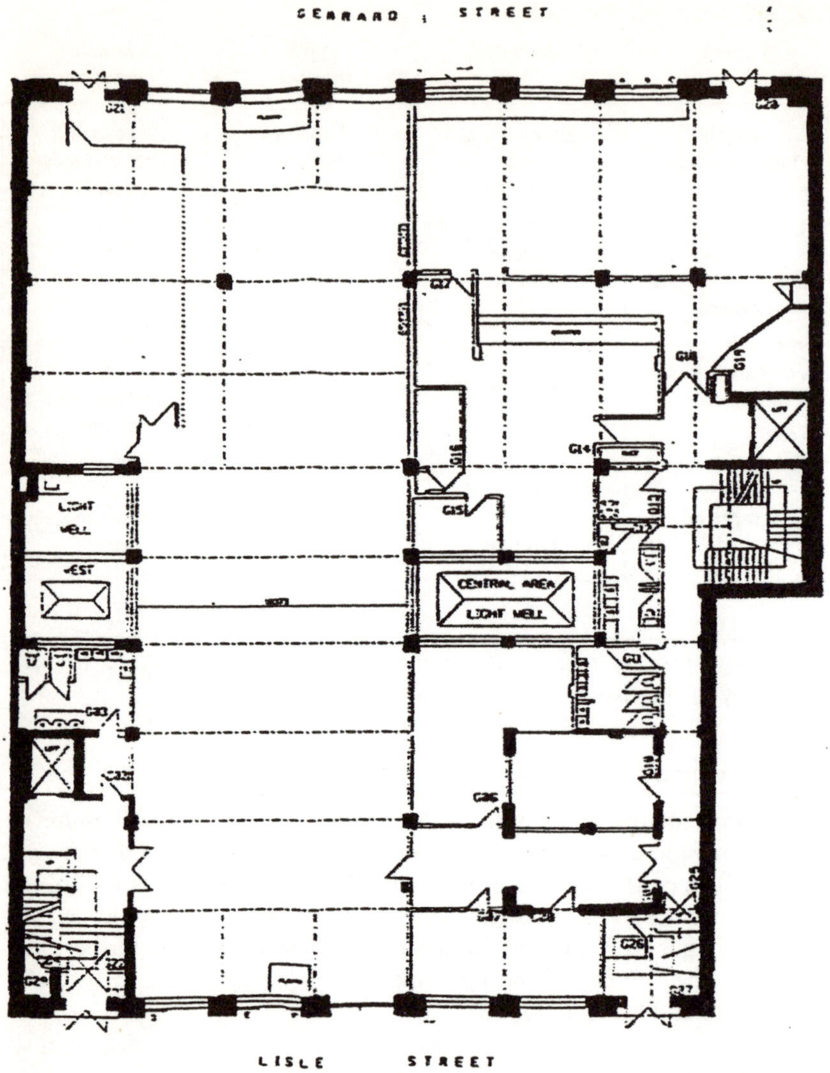

Figure 2.2 Gerrard Street Telephone Exchange – floor plan.

Figure 2.3 Gerrard Street Telephone Exchange – elevation.

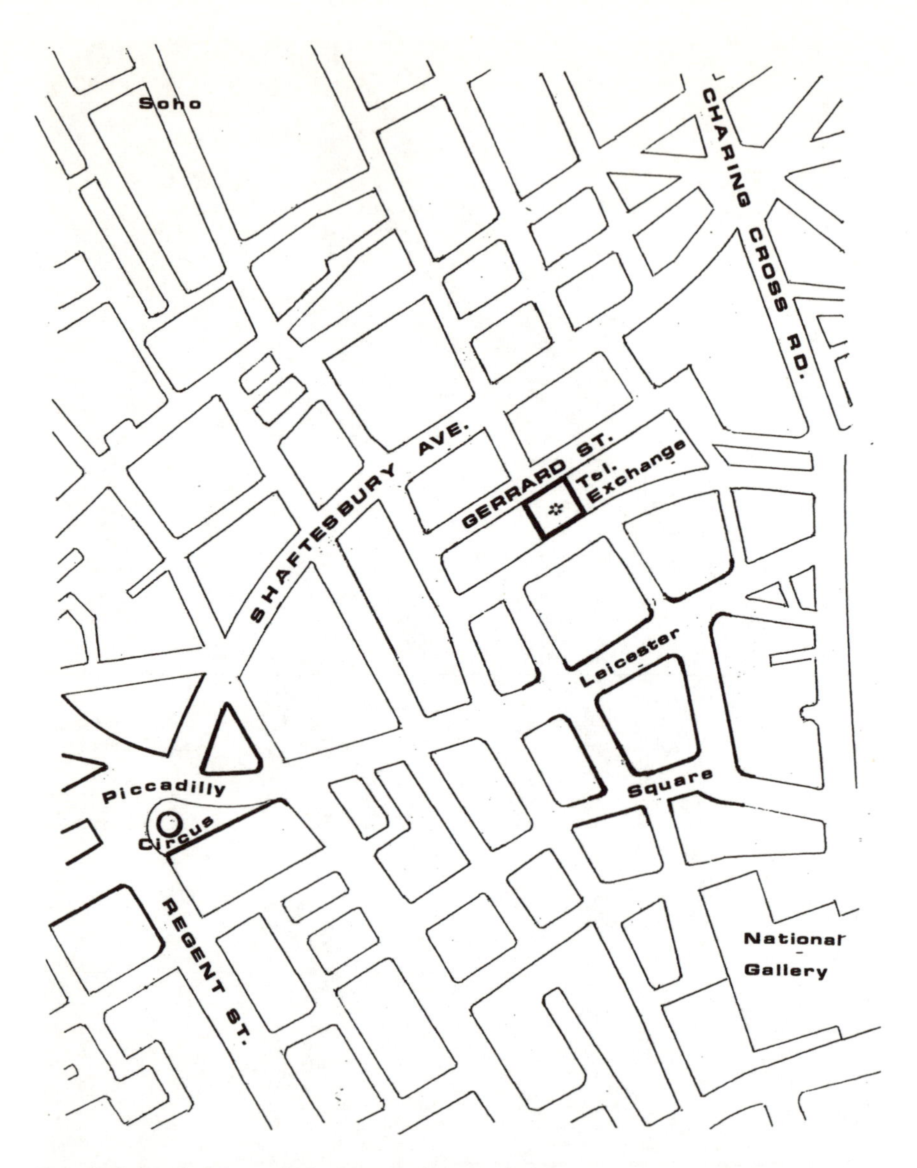

Figure 2.4 Gerrard Street Telephone Exchange – location.

3 Is the building available in whole or in part?
   - Gerrard Street TE has a very limited amount of equipment that must remain in its existing location within the basement of the building. The tenure rating is **partial.**

4 What is the existing floor to ceiling height?

- Gerrard Street TE has a variety of floor heights throughout the building. Upper floors up to 4.5 m, basement at 2.3 m. The majority of the floor to floor heights are in excess of 4.3 m. The floor to floor height range is **3.65 m to 4.65 m.**

5 What is the existing floor strength?

- Gerrard Street TE has a variety of floor strengths throughout the building. These were estimated to be between 8 and 10 kN/m. The floor strength is **5 kN/m to 10 kN/m.**

These first stage characteristics are illustrated together in Table 2.25, which presents a profile of Gerrard Street TE. These can now be used to compare the existing building against the requirements of the 77 uses in the comparator.

These first stage characteristics are compared and sorted against the 77 uses. When these are sorted in ascending order for excesses between the potential use and the existing building, the results are illustrated in Table 2.26. The 0 indicates that there are several potential alternative uses that might be considered. These can be more fully assessed by reviewing the second stage characteristics.

Second stage characteristics

The second stage characteristics deal with the key physical and locational elements of the building. The key physical characteristics are depicted in

Table 2.25 Gerrard Street TE: first stage characteristics.

UCO	*Hostile*	*Tenure*	*Height (m)*	*Strength (kN/m)*
B2–B7	Hostile	Whole	>4.65	>10
Other	**Not hostile**	**Partial**	**3.65–4.65**	**5–10**
			2.75–3.65	3–5
			2.3–2.75	<3

Table 2.26 Gerrard Street TE: top ten uses from first stage review.

	SIC use	*Excess use > supply*	*Matches*	*Shortfalls use < supply*
64.2	Telecommunications	**0**	5	0
52.1c	Retail sale in non-specialised stores, high spec (medium/small)	**0**	4	1
52.1a	Retail sale in non-specialised stores, medium/ small	**0**	4	1
93.01	Washing and dry cleaning of textile and fur products	**0**	4	1
52.2/4, 7	Retail sale in specialised stores	**0**	4	1
55.1/2b	Hotels, standard to luxury	**0**	3	2
65/70b	Finance, insurance and real estate industry, principal	**0**	3	2
55.1/2a	Hotels, low cost	**0**	3	2
52.6	Retail sale not in stores	**0**	3	2
55.3/5	Restaurants, bars, pubs, canteens	**0**	3	2

part through the photograph of the Gerrard Street elevation (Figure 2.3) and through the floor plan (Figure 2.2).

Physical characteristics

6 What is the existing building specification?
   - Gerrard Street TE has a very strong exterior facade that is more unique than standard. The stone arcade at ground level and first floor along the Gerrard Street facade are of much higher quality material than the surrounding buildings. The interior spaces are not of the same quality. The building specification rating is **unique exterior** with **standard interior.**

7 What is the existing building character?
   - Gerrard Street TE has a well-designed facade, the key element at ground level being the arcade. It has a strong presence along both streets and can be considered to have strong character. The existing interior spaces do not have any noteworthy features. However, that is not to say that the unusual light wells could not be used to advantage to create a unique interior. The building character rating is **strong facade** with **weak interior.**

8 What is the existing perimeter to perimeter depth of the floor plate?
   - Gerrard Street TE has a 33.24 m building depth at ground floor level (inside wall to inside wall from Gerrard Street to Lisle Street). The building depth is $\geqslant$ **24 m perimeter to perimeter** .

9 What is the existing building access?
   - Gerrard Street TE has four entrances at ground level, two for Gerrard Street and two for Lisle Street. It has two existing core areas. The building access rating is **multiple entrances** and **multiple cores.**

The physical characteristics are summarised in Table 2.27.

Locational aspects

10 What is the existing street characteristic?
   - Gerrard Street TE can best be described as highly integrated borough centre location. This is based on the fact that the city centre streets in the area are Oxford Street, Edgeware Road, Marylebone Road and Regent Street. The building street character rating is **highly integrated borough centre.**

11 What are the existing local amenities?
   - Gerrard Street TE is located on the edge of London's Soho district and near Leicester Square. This affords a variety of social–leisure amenities such as restaurants, bars, cinemas and access to squares. The social–retail amenities are also very strong. The physical vacancy amenity is strong, as there are very few vacancies or signs of derelict buildings in the immediate area. The building local amenities rating is **strong social** and **strong physical.**

Table 2.27 Gerrard Street TE: second stage characteristics.

Specification	*Character*	*Depth*	*Access*
Exterior/interior unique quality	Strong facade and interior	**>24 m p to p**	**Multiple external and cores**
Exterior unique quality	**Strong facade**	24 m p to p	Multiple cores
Interior unique quality	Weak facade	18 m p to p	Multiple exterior
Exterior/interior standard quality	Weak facade and interior	≤12 m p to p	Single external and cores

12 What is the existing public transport provision?
  - Gerrard Street TE has access to two Underground stations within 5 minutes and a variety of buses that stop along either Shaftesbury Avenue or Charing Cross Road. There is no train access within 5 minutes. The building public transport rating is **two forms** within five minutes.

13 What is the private transport provision?
  - Gerrard Street TE is accessible to either Shaftesbury Avenue (A401) or Charing Cross Road (A400). The building private transport rating is **A road ready access.**

The top uses were re-examined for matches following the second stage (Table 2.28). The top ten uses indicate some of the potential uses that could be considered for the building as it currently exists. It is notable that the inclusion of the detailed physical and locational characteristics has given emphasis to a more interesting set of specific options, such as luxury hotels and finance offices, while retaining the more obvious retail choices. Buildings in less central locations were found to suggest an even wider range of choices from the 77 options (Table 2.29).

Use of the comparator to aid decisions

The Use Comparator developed during the course of the research is now available to assist others in finding viable uses for existing buildings. The

Table 2.28 Gerrard Street TE: locational aspects.

Street	*Amenity*	*Public transport*	*Private transport*
Highly integrated/city centre	**Strong social and physical**	**3** forms within 5 minutes	A road and airport-ready access
Highly integrated/borough centre	Strong social	**2 forms within 5 minutes**	**A road-ready access**
Partially integrated	Strong physical	**1** form within 5 minutes	A road-limited access
Poorly integrated	Weak social and physical	**0** forms within 5 minutes	A road-remote access

Table 2.29 Gerrard Street TE: top ten uses after second stage review.

	SIC use	*Excess use > supply*	*Matches*	*Shortfalls use < supply*
55.1/2b	Hotels, standard to luxury	0	**5**	3
52.1c	Retail sale in non-specialised stores, high spec (medium/small)	1	**4**	3
65/70b	Finance, insurance and real estate industry, principal	0	**3**	5
52.1a	Retail sale in non-specialised stores, medium/small	0	**2**	6
55.1/2a	Hotels, low cost	0	**2**	6
64.2	Telecommunications	0	**1**	7
52.6	Retail sale not in stores	0	**1**	7
55.3/5	Restaurants, bars, pubs, canteens	0	**1**	7
74	General business activities and services	0	**1**	7
75	Public administration and defence; compulsory social security	0	**1**	7

customised spreadsheet which includes all of the use characteristic scoring for the 77 uses in the new classification framework is held at the Bartlett School of Graduate Studies at University College London. In order to identify a list of potential uses for a building, it is of course necessary to score the characteristics of the building and its location and then to operate the Use Comparator program. The list of options that emerge from this then have to be evaluated financially given prevailing market conditions before any final decisions can be taken. The Bartlett will assist in scoring a building and will operate the program for a small fee.

Iterations on the original decision

Following the original research investigations, an additional related investigation considered the effects and extent of what was called selective demolition in refurbishment work. As a result of this work it became apparent that sometimes a significant range of new possible uses could be developed for a building through such demolition activities. Thus moving an entrance from one face of a building to an opposite or adjacent face could change buildings both physically and, in part, locationally. The Use Comparator can be used very simply to consider the general effect of these kinds of changes by simply readjusting the score for the building to reflect these possible changes. Thus it provides a mechanism for testing several iterations of the decisions process with very little demand on resources. This is discussed further in section 2.4.

Key comparator features – a summary

- The Use Comparator is a two-stage decision aid which can suggest the most appropriate uses for a redundant building with given characteristics in a particular location.
- Thirteen physical and locational characteristics are measured on various scales for each of the 76 uses and these are compared with these same characteristics for the building available.
- Potentially valuable additional uses can be found by testing selective demolition options which would modify the subject building.
- The comparator system is available at the Bartlett School of Graduate Studies for use on a fee basis.

2.4 Physical change possibilities

Exploring the extent and nature of physical change

In considering physical change possibilities, it is necessary to be able to differentiate between the relative ease or difficulty of changing any particular physical feature of an existing building. At one extreme, some physical characteristics can be easily and cheaply modified and improved through minor works alone. Other factors can only be changed with substantial adjustment and reconstruction. At the other extreme, the adaptation of many physical characteristics is impossible without major structural alteration, often involving a degree of demolition. Furthermore, while many physical attributes of a building can be completely changed during adaptation, some physical characteristics can only be qualitatively improved and enhanced. The means for profiling the physical characteristics of a redundant building were described in section 2.1 and outlined in Figure 2.1. In this section these characteristics will be further examined to identify the nature and range of what has been described as selective demolition.

In evaluating the general viability of a 'change of use' development, it is essential first to review the scale of physical intervention that might be appropriate to the specific project circumstances. This involves the examination of:

- new uses that are viable with negligible physical change to the building 'as found'

- new building uses that could be viable with minor physical change
- uses that might become viable, given significant physical change, reconfiguration and reconstruction.

Two types of physical change must be considered: those to the external fabric of the building, and those to the internal spaces and layout. During the refurbishment process, the existing external fabric of a building can either be maintained in a largely unchanged state, or modified and upgraded. Alternatively it may be completely replaced with recladding, re-roofing and new fenestration. In a similar way, internal changes may be minor, with adjustments to space and finishes only, or major, involving significant structural change and complete spatial reconfiguration. The internal and external change characteristics give rise to four basic strategies for adaptation, as illustrated in Figure 2.5. These are as follows.

- LOW CHANGE: maintain the existing external fabric with minor modification of the internal space.
- LOW–MEDIUM CHANGE: replace the external fabric and modify the internal space with no structural change.
- MEDIUM–HIGH CHANGE: maintain the external fabric, reconfigure the internal space, with some modification of the building's structure.
- HIGH CHANGE: replace the external fabric, modify the building's structure and reconfigure its internal space.

	MAINTAIN EXTERNAL FABRIC	**REPLACE EXTERNAL FABRIC**
MODIFICATION : INTERNAL SPACE only	Low change	Low–medium change
RECONFIGURATION : SPACE & STRUCTURE	Medium–high change	High change

Figure 2.5 Types of physical change.

Reviewing the options for change

The basic options for physical change can be systematically explored through a simple five-stage process to:

- identify the range of options that are available for physical change within one of the four 'low to high change' strategies, as defined in Figure 2.5
- eliminate any options that are not technically viable or practicable for the particular building under consideration
- eliminate those physical change options that are not compatible with the potentially viable new uses as identified by the Use Comparator procedure
- select two to five preferred options from the set of potentially viable physical adaptations for detailed consideration
- conduct comparative business case and technical design evaluations, and select the preferred scheme for implementation.

The extent, type and combinations of physical changes that are to be undertaken are of fundamental importance to any refurbishment strategy overall. Figure 2.6 sets out the basic development combinations. If a 'low change' strategy involving minor physical change is to be adopted, then its viability will rest mainly on the flexibility of the building 'as found', and a low-cost specification (areas 1 and 2 in Figure 2.6). The class of future use is the most critical single factor affecting the specification of materials and finishes for a refurbishment project, and the resulting costs of adaptation. Survey results indicated that project specification tended to be 'supply' rather than 'demand' led, with the level of specification dependent on the client's brief (80%), project budget (75%), and design and developer criteria (68%) rather than user demand (50%), location (34%) and target price (31%). In summary, a 'low change' strategy will depend on:

- the morphology and dimensions of the building, its floor plate, structural grid, floor to floor height and fenestration modules being suitable, without change, to support the proposed new use
- the overall flexibility of the existing building space being sufficient to permit replanning and redesign for the new use
- the construction and materials of the building 'as found' being appropriate for the proposed new use
- the feasibility of adopting an appropriate 'low change' specification strategy for services, fittings and finishes.

In the case of 'high change' refurbishments, it became evident from the investigations undertaken during the research that extensions to the existing building, together with some partial demolition, often formed an important part of the development strategy (areas 4, 5 and 6 in Figure 2.6). Field

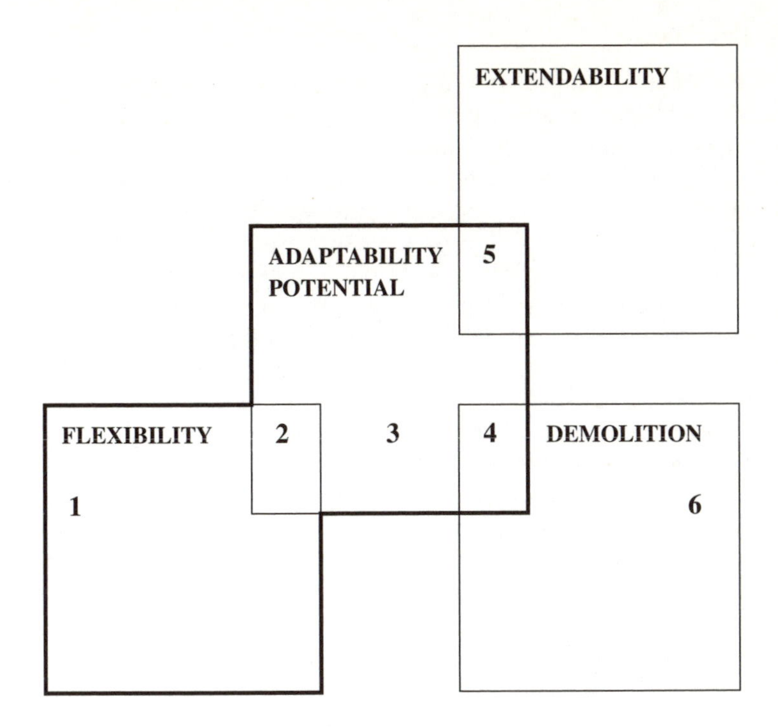

1 **Change of use through flexibility of the building 'as found'**
2 **Change of use through flexibility with minor adaptation**
3 **Change of use adaptation/refurbishment of vacant facility**
4 **Change of use adaptation with selective demolition**
5 **Change of use adaptation with extension of facility**
6 **Change of use through demolition and redevelopment**

Figure 2.6 Basic development combinations.

investigations showed that both horizontal and vertical 'new-build' extensions were common. It was also evident that a degree of partial demolition frequently played a critical part in development strategies, generating additional options for adaptive reuse. While the surveys showed that there was a general awareness of the crucial role that new-build extensions can play in the refurbishment of buildings for new uses, there was little recognition of the potential value of partial demolition to provide additional opportunities for the reuse and mixed use of existing buildings. Decision agents were generally unaware of the potential for expanding 'change of use' options and enhancing value through a degree of selective demolition. The issues and opportunities of selective demolition appear to have been neglected in refurbishment research and best practice advice.

Selective demolition

Selective demolition is defined as the conscious removal of some parts of a building's usable floor space, in addition to the demolition during refurbishment of specific elements of buildings such as walls, services and parts of primary structure. Selective demolition can be part of two of the four main strategies for physical adaptation, as shown shaded in Figure 2.7. A five-part scale is also shown, indicating the range from 'low' (<20%) to 'high' (>80%) amounts of selective demolition. The extreme situations of no demolition (0%) and total demolition (100%) set the theoretical bounds within which redevelopment proposals can be considered and analysed. This provides the basic scale against which the impacts and benefits of different degrees of selective demolition can be compared. By definition, refurbishments incorporating selective demolition would generally result in a reduction in the overall quantity of usable floor space. From a commercial standpoint this reduction will need to be compensated by an increase in the 'use value', 'rent value' and/or 'asset value' under the building's new uses.

Case investigations were undertaken of current and recent 'change of use' projects that involved selective demolition to:

- identify additional options for 'change of use' that can be achieved through a degree of selective demolition
- gain an understanding of the potential impacts and benefits of different types of selective demolition
- alert decision agents involved in refurbishment to the opportunities of selective demolition.

Quantitative data were extracted from planning application records and archive documents. Qualitative information was obtained from structured interviews of the decision agents involved. These data were analysed to compare systematically the situations 'before' and 'after' refurbishment,

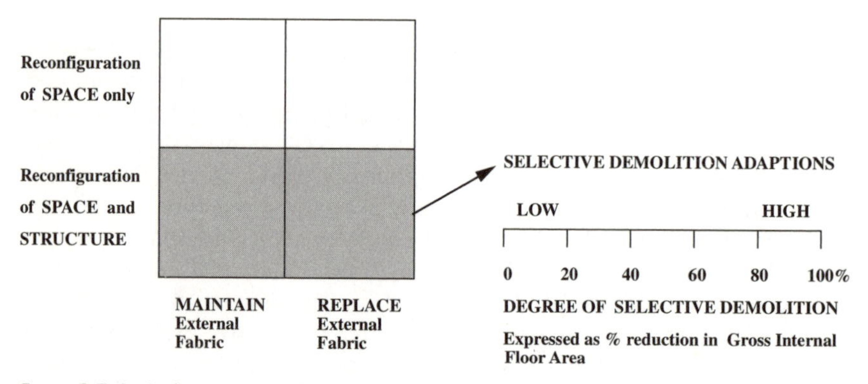

Figure 2.7 Scale for selective demolition.

and to assess the impact of selective demolition on the adaptive reuse of buildings generally, identifying:

- physical attributes that had been modified by selective demolition
- physical characteristics that had been enhanced by selective demolition
- the ways in which selective demolition had affected options and outcomes
- decision agent perceptions of the critical issues involved
- the relative impact of selective demolition on 'change of use' viabilities generally.

Physical characteristics changed in selective demolition

Of the original 25 physical characteristics identified in section 2.1, 18 were significantly changed or modified as a result of selective demolition. These are shown in Figure 2.8 In cases where considerable amounts of demolition had occurred, gross internal areas were largely unaltered before and after refurbishment. Planners had tended to allow extensions and additional floors to compensate for space lost by selective demolition, achieving an approximate balance overall. Selective demolition was a particularly essential component in two particular types of development:

- changes to deep-plan office and warehouse buildings to residential use, commonly requiring the reconfiguration of core areas
- changes from 'single' to 'mixed' use developments, involving new internal circulation routes, multiple entrance conditions and new compartmentalisation (horizontal and/or vertical) of the internal space.

The most common changes achieved through selective demolition are listed below.

Site

- Site access was significantly changed as a result of selective demolition with additions to both pedestrian and vehicle access.
- Parking provision was significantly increased as a result of selective demolition and this was seen by decision agents as an important factor in increasing property value.
- Site density (plot ratio) was only slightly modified in all but one case. It would appear that in residential adaptations a higher density (habitable rooms per hectare) had been achieved than would normally be permitted.
- Site amenity was positively affected by selective demolition in 80% of cases.

Physical attributes				Potential changes
Site	Site access			●
Site	Boundary conditions			
Site	Density			●
Site	Parking provision			●
Site	Site amenity			●
Space	Size	Total floor area		●
Space	Size	No. of storeys		●
Space	Size	Floor plate areas		●
Space	Size	Primary units of space		●
Space	Shape	Depth p/p		●
Space	Shape	Depth p/c		●
Space	Shape	Length		
Space	Linkage	Internal	Vertical	●
Space	Linkage	Internal	Horizontal	●
Space	Linkage	External	Entrances	●
Space	Linkage	External	Fire escape	●
Space	Linkage	External	Party walls	
Fabric & structure	Character	Facade		●
Fabric & structure	Character	Interiors		●
Fabric & structure	Strength	Foundations		
Fabric & structure	Strength	Superstructure		
Fabric & structure	Dimensions	Bay width		
Fabric & structure	Dimensions	Bay depth		
Fabric & structure	Dimensions	Fl to fl height		●
Fabric & structure	Dimensions	Elevations		●

Note: ● attributes typically changed by selective demolition

Figure 2.8 Physical characteristics changed by selective demolition.

- The removal of poor-standard extensions and elements of existing building complexes resulted in the introduction of uses requiring high visual quality overall.

Space

- Total floor areas were subject to marginal changes overall, both in gross internal area and in net usable area. Where substantial amounts of demolition were undertaken (more than 20%), this tended to be compensated by adding floors and horizontal extensions.

- The number of storeys was increased in 30% of cases.
- Primary units of space (undifferentiated space within the building) were significantly changed by selective demolition, with the reduction in depth being seen as a major factor permitting change of use.
- Depths perimeter to perimeter and perimeter to core were changed, with particular relevancy to residential reuse.
- There would appear to be a substantial level of demolition of core areas in refurbishments for change of use. Vertical linkages were changed in all of the cases considered. The degree to which core areas were demolished and repositioned, or demolished and reconfigured, was a major finding of the extension study.
- Horizontal internal linkage routes were changed in 60% of cases to provide new internal access routes and/or fire escape routes. These changes were of critical importance in adaptations to residential use.
- Horizontal external linkages were modified through selective demolition in 80% of cases to achieve new entrance positions and external fire escapes.

Fabric and structure

- The character of the facade was substantially changed in relation to cladding–glazing ratio in most cases.
- The character of the interior was modified by selective demolition in nearly all cases.
- Floor to floor height dimensions were modified in 30% of cases.
- The dimensions of elevations had been materially changed in most cases, as would be expected in major 'change of use' refurbishments.
- In all of the cases investigated, there were no changes to the basic structural bay widths and depths of the building as a result of selective demolition, indicating that these might be considered an invariant feature in most 'change of use' refurbishment.

The work undertaken has led to a clearer understanding of the generic opportunities for reuse afforded by adaptations involving different degrees of selective demolition, over and above those that can be realised through basic refurbishment alone. These results consistently confirmed that a degree of partial demolition is beneficial, first to extend the range of possible uses of redundant buildings, second to achieve environmental improvement and energy savings, and third to introduce new financially viable options for use. Here the increase in net lettable area, increased parking provision, and physical improvements to the character of the facade were seen to be the factors having greatest positive effect on capital values, rental rates and market price. This work has established the main reasons for undertaking selective demolition adaptations; it has identified the range of options explored, the criteria used for project appraisal and the expected and realised benefits.

Summary – the benefits of selective demolition

The major benefits that can be achieved from these physical change possibilities include the following.

- The opening up of *deep-plan space* through selective demolition of floor plates to introduce atria, light wells, and interior streets to achieve high-quality space suitable for housing, multiple retail, other multiple uses and sub-letting and all uses requiring shallow-depth floor plates.
- The creation of new entrance arrangements, parking solutions and public access arrangements at ground and lower ground levels, to permit the introduction of *public uses* and *multiple tenancies* for small-scale retail, recreational, cultural and social uses.
- The introduction of new vertical and horizontal circulation elements, and party wall and floor divisions, through which major reconfiguration of large items of building stock may be achieved to permit change from *single* to *multiple* or *mixed uses*.
- *Core*: A substantial amount of selective demolition to the core of buildings is frequently carried out as part of refurbishments for 'change of use'. The adaptation of parts of buildings that hitherto had been considered as fixed and invariant had opened up new opportunities for reuse. This finding suggests the need for a fundamental reconsideration of traditional design concepts and that the flexibility and adaptability potential of buildings relates, in the main, to the possibilities of reconfiguring usable floor space. The results suggest that consideration, during new-build design, should be given to the opportunities for the radical modification of the spatial morphology of buildings to permit reuse and mixed use in the future.
- The removal of parts of building floors to relax the *floor to floor height* constraints, particularly at ground and first floor levels, permitting the introduction of otherwise non-viable uses such as retail and recreational uses.
- The introduction of daylight and natural ventilation into existing buildings, reducing or eliminating their reliance on AC systems, and improving end-user *comfort* and *energy performance* for new uses that are normally unviable in fully conditioned environments, particularly in relation to residential uses.
- The opportunity for the development of high 'use value' spaces within existing buildings of *character* or *historic importance*, conserving

quality elevations and interiors, retaining features of historic importance, while refurbishing or introducing space that is appropriate for contemporary uses.
- *Planning benefits* and *gains* through changes that improve site density, site amenity and parking provision, and that assist planning authorities in meeting their planning objectives, particularly those under the SAGE (Sustainable Action for a Greener Environment) programme, their policies to encourage mixed-use development, and their targets for social housing provision, mixed residential unit developments and non-family housing generally.

Having outlined a number of ideas on how to view both the supply and the demand side of adaptive reuse and described a mechanism for comparing the two and a framework for considering the selective demolition of an existing building, there remains one crucial issue to be considered. This is the question of viability of the proposal in relation to the circumstances prevailing at the time of the decision. Though this is a mature subject on which much has been written in the past, in the following section a few fresh ideas are developed to complete this chapter.

2.5 Assessing viability

Aspects of viability

The procedures for investigating the basic families and combinations of the physical and use options that are potentially available to those undertaking refurbishment have been identified and mapped. A systematic search procedure has been outlined, through which non-viable options are discarded at each stage of the process, converging towards a small and secure subset of adaptation possibilities. This small set of viable and preferred adaptation possibilities needs then to be subjected to detailed analysis, comparison, and final selection of the 'best' option. The basic decision framework for the review of project viability overall is illustrated in Figure 2.9, showing the three key areas of decision relating to:

- **use viability** – the identification and assessment of use possibilities, within the regulatory framework of planning and heritage legislation
- **technical viability** – the investigation of physical change possibilities, and the overall adaptability potential of a building, within the constraints of the building regulations and other relevant legislation
- **financial viability** – the financial appraisal and comparison of 'change of use' development options, within the context of the prevailing market conditions.

In regard to the first two areas in Figure 2.9, the procedures for identifying change of use and physical change possibilities have been described in sections 2.3 and 2.4 respectively. In the third area shown, that relating to financial viability, there are a number of well-established and proven methods for the financial appraisal of development projects, such as are most interestingly presented by Eley and Worthington's (1990) report[24]. In

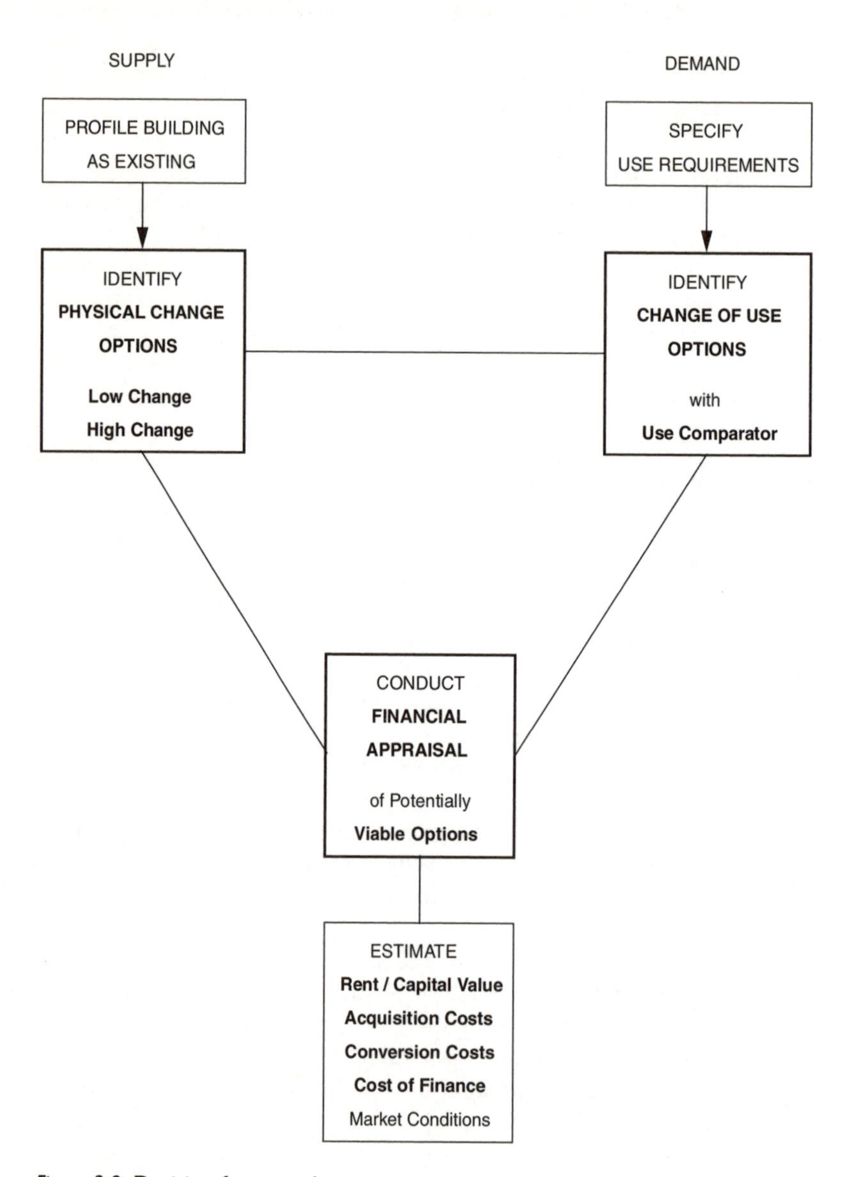

Figure 2.9 Decision framework.

fact, this area is very professionally covered both within and without the subject of buildings and facilities, therefore this volume will not attempt to cover this ground again. However, the reader may find the commentary in Chapter 3 on investor sources and returns in refurbishment projects of some value in setting the parameters for financial analysis.

Sequence of analysis

The research results indicate that there is no single preferred sequence in which these three problem aspects should be addressed. A variety of alternative sequences can be adopted, depending on the specific circumstances of a particular project. In the majority of cases, however, the process will begin with a vacant building that appears to be redundant under its current class of use. A search for new uses that are potentially viable with minimum physical change to the building will then be undertaken, as described above. If new uses with minimum physical change are identified by the Use Comparator system, then these will be examined for general financial viability. Should no or very few options be identified, then an iterative process of further consideration will need to be undertaken to determine whether viable uses are uncovered, but with significant degrees of physical change to the building. Again, an approximate financial appraisal of the options will need to be undertaken during each cycle of iteration. Once clear options have been identified that appear to be viable on 'use', 'technical' and 'financial' grounds, then two or three preferred options should be selected for detailed evaluation, comparison and development.

A variety of alternative cycles of iteration and sequences of decision are available to meet the specific requirements of a particular project. Figure 2.10 indicates the two basic sets of possibilities: a 'supply-led' approach as outlined above, and a 'demand-led' process for circumstances where the

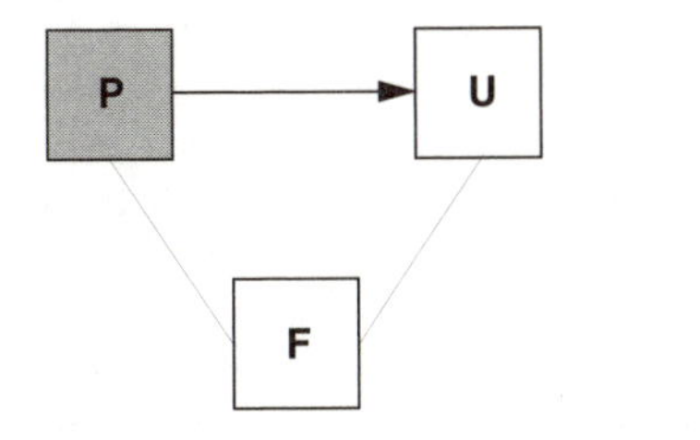

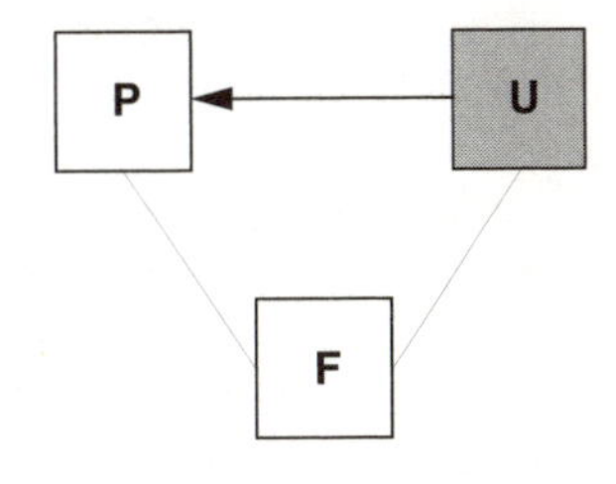

SUPPLY-LED

- **Identify Potential Uses for any given Redundant Building**

DEMAND - LED

- **Identify Potential Buildings for any stated Type of Use**

Figure 2.10 Supply-led and demand-led approaches.

search is to identify potential buildings for change of use adaptation to meet the requirements of a given client or stated class of use.

A 'finance-led' approach is probably not uncommon, particularly from investors/developers who specialise in adaptive reuse refurbishments and who seek either to meet market demand (conversions of derelict industrial properties to housing near city centres are an example) or to exploit supply opportunities. In addition to changing the starting point for such decisions, it is useful to consider the direction of the first iteration going from P (the supply) to F (finance) or U (demand) as circumstances change or opportunities are perceived.

Potentials for failure

In this section most of the commentary has related to various aspects of the physical and locational make-up of buildings. The ideas outlined assume that all of the skills are in place to understand and implement appropriate standards of quality and cost sufficiently to meet functional and market expectations. Viability depends on this, but it also depends on another set of skills which the research showed do not necessarily exist with every member of the coalition of specialists required to effect adaptive reuse. As illustrated in Figure 1.4, not all decision agents look at viability with an equal balance of emphasis on cost, value, risk and robustness. In particular there is too little emphasis on value in the producer group, though there is a strong emphasis on cost. Thus the 'hard data' that can be used in the cost and physical measurements side of an assessment have to sit beside the 'soft data' that are assumed for market value and for some aspects of building characteristics such as amenity.

However, this implies that the systematic approaches that are described in this section are arguably even more important in decision-making than they would be if all the data were as hard as that which we can obtain for costs or structural strengths and building dimensions. Inevitably, after all of the characteristics of supply and demand have been assessed and valued, compared to each other and the investment options analysed, there will remain areas of judgement to be considered. These judgements may be even more crucial than the measurements and values, but the risk of project failure is reduced even then by using a clear and logical method informed by contemporary research. When only overall judgement or inspiration is used, as was the case in too many of the refurbishments surveyed, then the risk of failure is significant.

Summary

Five fundamental questions have been addressed in this section as follows.

- What are the key factors that can affect 'change of use' refurbishment decisions?
- What set of characteristics can render a given building 'more adaptable' or 'less adaptable' within its general class of use?
- What 'supply-side' data are needed to support an assessment of the adaptability potential of a given item of stock?
- How can potentially viable uses be identified?
- What decision framework and procedures might be adopted for the assessment of the use viability, technical viability and financial viability of a 'change of use' proposal?

Having decided on the most appropriate new use for an adapted building, there remains the considerable task of delivering the design and construction of the project. Throughout the research it became apparent that there were a number of management issues for which the results were relevant. Chapter 3 deals with many of these. It is not intended to be a primer on project management, about which much excellent work has already been done, but rather a set of insights into some of the specific project management issues one can expect to encounter in adaptive reuse.

Chapter 3

Securing the management of implementation

It is well beyond the scope of these guidelines to attempt to summarise what has been so completely covered in other literature about project management, whether for refurbishment or for new buildings. However, it is evident from the research that there are aspects of management associated with adaptive reuse which are possibly unique or are at least of special importance to this activity. This chapter is therefore an attempt to draw attention to certain aspects of financing, human resourcing, decision-making, project organisation, marketing and prioritising which, if handled with some knowledge of the special characteristics of adaptive reuse, can give greater security to management within this specialist area.

In each of the following sections a summary of the key points is presented at the beginning to allow the reader to identify quickly those aspects of the activity which are covered and may be worthy of further investigation. The text following the summary is based on the findings of the research and is there to substantiate and elaborate on the points made in the summaries. Section 3.1 deals with the cost and sources of finance and the opportunities for investors as sustainability concepts encourage more adaptive reuse activity. Section 3.2 considers the experience and skill needed to carry out the roles of the five principal decision agents involved in this activity. Section 3.3 identifies key decisions and discusses how these affect major risks and the long-term robustness of projects. Section 3.4 deals with a number of issues associated with planning approval, the working arrangements of designers and contractors, and the attitudes and preferences of building users.

3.1 Investment finance for adaptive reuse

Key investment issues

- Investment in adaptive reuse yields 0.5% to 1.5% more than new build.
- Investors focus on ability to repay in preference to quality of project.
- Investment portfolio diversification is away from property and towards bonds or foreign equities.
- Investors in adaptive reuse are typically relatively small specialist developers, or foreign financial bodies.
- Increasing interest in sustainable development alongside rapid technological change suggests investment growth in adaptive reuse.
- Project loan size and risk can be reduced by phasing projects to assist early income and to align design solutions with market preferences.

Lending costs for refurbishment

Lenders such as banks, insurance companies and pension funds distinguish between change of use refurbishment and new build, and clearly prefer the latter. Thus to sell equity in the adapted building or to obtain construction loans for adaptive reuse is even more difficult than the already demanding task of raising money for new property investments, which at the time of writing remains a relatively unpopular avenue for investment. Thus our surveys showed that a clear majority of respondent lenders distinguished between refurbishment and new build by ranking the former as a higher-level investment risk which is assumed to yield close to 0.5% more. Actual yield at any given time is linked to returns available at that time from equities and is also dependent on the use category, as illustrated in Table 3.1 for a particular period in 1995. It is important to note that most surveyors would advise that the yield spread between new and refurbished is closer to 1.5% historically.

Investment diversification away from property

In the case of project loans, it is important to note that the researchers were clearly told by **lenders** that they **have little interest in the project but**

Table 3.1 Investor group's anticipated yield performance by sector.

Sector	*Yield mean new build*	*Yield range new build*	*Yield mean refurbishment*	*Yield range refurbishment*
Retail	7.6%	6–10	7.8%	6–10
Office	8.0%	6–12	8.6%	6.5–12
Industrial	9.6%	8–14	10.2%	9–14
Residential	8.0%	8	8.0%	8
Other				

rather focus on the ability of the borrowing organisation **to repay the loan.** In this regard it is worth noting that banks regard their legal advisers as their most important professional advisers for project loans and only by exception refer to other professionals such as property surveyors or architects.

The general difficulty in property investment arises from the wide range of choices open to institutional investors in domestic and international equities and debt where transaction costs are low and times are short, while returns are good. Long-term investment is therefore available largely from limited portfolio diversifications in pensions and insurance for UK companies and from foreign investors seeking to participate in UK property investment from time to time. Investment fashions change like all others, but the structure of investment under current GATT agreements favours diversification through external investments. In the housing sector, unless and until legislation is clearly seen to encourage the return of private sector savings, it is unlikely that investors will show much interest in refurbishment opportunities. The withdrawal of long-term investment by pension funds and insurance companies from the multiple-dwelling private housing sector in the UK in the three decades from 1950 to 1980 is not likely to be quickly reversed, as it represented a major loss of confidence in this sector. Accordingly it should be no surprise to discover that, with the exception of a few large-scale project loans, the principal investors in adaptive reuse are relatively small specialist investors who see the potential for specific opportunities to be exploited. A useful source of information on who these organisations are is the *UK Directory of Property Developers, Investors and Financiers*, published by Spon Press.

Investment opportunities in sustainability

Investors looking at new opportunities will inevitably remain concerned about the high transaction costs and slow pace of adaptive reuse. However, there are some substantial reasons why it might be worth looking past these obstacles to see the opportunities within this area of activity. Much has been said and written elsewhere about the extraordinary pace of change and competitiveness of the world economy as factors such as the spread of information technology and the liberalisation of trade agreements create new ways of doing things and accelerate obsolescence of much of our economic infrastructure. While this is happening, concern increases about the impact of human activity on the physical environment and governments increasingly seek to restrain and shape the choices that individuals and organisations make. These two fundamental forces converge in the way our cities are changing and developing, and increasingly policy-makers speak of concepts such as sustainable development as a way of characterising an enlightened approach to accommodating these forces. This is discussed further in Chapter 4.

Adaptive reuse of offices and industrial buildings

In London, as in many other cities, the convergence of global and environmental forces has been seen in the periodic creation of significant surpluses of office space as large users such as oil companies, telephone companies, computer companies, government and others find that IT significantly reduces the numbers of administrative and professional staff required to do the same or more work. IT also enables organisations to provide less space for the same number of people through concepts such as hot-desking. Whole buildings are often vacated and, unless they are top category, they often become candidates for change of use and in the dominant number of cases (see Figure 1.2) are refurbished for housing use. As mentioned before, this phenomenon is discussed in detail in the Home Office Report[25] for those wishing to understand better the scale of this activity. The effect of these adaptations is to bring people closer to their work and to revitalise city centres as people start to return from the suburbs to areas which had become dominated by office uses. The environmental benefits are obvious, as, perhaps, are the investment opportunities.

The de-industrialisation of major cities has an even longer history and, because of the largely unsuccessful attempts by town planners to protect the employment capacity of buildings, has often led to more obvious dereliction in parts of the urban landscape. Thus the move from industrial use to housing or retail has been slower despite the emergence of warehouse apartments in very conspicuous locations such as river frontages. Often large-scale investments are required here, as was the case for instance for Butler's Wharf, adjacent to Tower Bridge in London. However, with some care to manage factors such as the size of increments of development and the use of appropriate teams of designers and constructors, these represent secure long-term opportunities that connect directly to the development and redevelopment of the cities which are at the centre of our economic life. Thus adaptive reuse of buildings is not a marginal activity but rather is as central to the renewal and change of cities as is new building. Therein should lie appeal to investors interested in opportunities to fund the projects or to invest in the long-term performance of the built environment.

Public funding of social housing

These guidelines are not intended as a source for information on the special funding that is available to housing associations through the allocation system operated by the Housing Corporation on behalf of central government in the UK. These allocations, which are based on needs indices which are derived from scoring a wide range of social disadvantages suffered by individuals and families, provide funding for investment in both new housing and adaptive reuse. Thus this is a source of low-cost capital and revenue supplements for housing which historically develops from its social role into the mainstream of commercially available housing as ownership transfers. Accordingly, although it is part of the market for funding and sales, it is mainly of interest here in relation to its participation in the conversion of traditional properties into housing from commercial use. The Housing Corporation assumes that these conversions cost more than new builds and provides funding accordingly.

Borrowing for projects

It has already been said that our evidence shows that the project loan decisions are based on the ability to repay rather more than on the project. We also found that the majority of these loans are for less than 80%, often no more than 50%, of the project value. This can create problems for borrowers with limited security or status but with a promising project and the necessary skills. To overcome this it may be necessary to limit the amount borrowed by breaking the project into phases, such that income generation or sales feeds funding back into the project as the project proceeds. This also has the advantage of testing the market for the designs being produced so that changes can be made as problems arise or opportunities emerge. Particularly in housing schemes this strategy can also be aligned with a pricing policy which works backwards from what is affordable to buyers into what should be designed and built in the scheme. Some of the industrial conversions use this approach in the context of selling unfitted space rather than complete apartments. All this may seem fairly obvious, but it was noted that some of the higher-cost schemes foundered financially because they chose to take a more monolithic approach. This is discussed further in section 3.3. It should also be noted that close to 80% of lenders have formal procedures established for evaluating risk on project loans. In these procedures, in addition to evaluating the borrower, they use internal or external staff to look at factors such as project location, the nature of the project, market norms and the return on investment. It is also usual for lenders to require that you have planning permission, retain professional advisers and have clarity on land tenure as conditions of any loan.

3.2 Assembling the skills and experience

Key skills and experience issues

- For adaptive reuse, designers have to be more innovative in the use of space and constructors more inventive with methods.
- Architects use what they find to create often more dramatic and unique spaces than are found in new buildings.
- Engineering consultants should have experience with the unique type of work required for each project.
- Choose contractors who specialise in refurbishment and who suit the use intended use and specification required.
- Marketeers have a critical role to play in change of use decisions affecting finance, design, pricing and timing.
- At every stage of scheme development, planners and other regulators should be included and their advice sought.
- More could be done directly by developers, marketeers and designers through market surveys and stated preference surveys to understand what users want.

Special qualities needed for adaptive reuse

If the property and construction industries attract a special breed of people who prefer informal organisations, deal-making, irregular hours and career discontinuities, then adaptive reuse adds a few characteristics of its own to that list. Designers have to be even more innovative in finding unique solutions to create new uses for space intended for other things. Constructors need to be more inventive in developing special methods to hold together some building parts while others are demolished around them. Developers acting as or for investors not only have to find a suitable building ahead of their competitors, but also have to see the market potential for a use and the market potential beyond that if the first use fails. Marketeers have to go beyond simple comparisons of similar properties and look to the subtleties of potential user demands. Regulators have to be drawn into the dialogue from the start to help shape the project to suit both the needs of the community and the realities of project economics and potential user requirements. The research confirmed that while the above-listed qualities are perhaps not surprising, not all members of the team adequately understood their importance. The following sections will deal in more detail with some of the evidence for this, and suggest remedies where possible.

Architectural design opportunities within apparent constraints

It was evident from our case studies that designers, particularly architects, were much less constrained by the characteristics of the adapted building as found than might have been assumed by casual observation. In all cases, of course, there was at least one major constraint to be found in each building, be it facade, structure or floor plate size and shape, but often not more than one. Thus designers moved entranceways from one facade to another, relocated staircases quite freely and generally found that they needed to pay scant regard to the basic building configurations deriving from the earlier use. Often designers make use of what they find to create more dramatic and unique spaces than might be experienced in conventional buildings. This is particularly true of adaptations to housing from industrial uses. Very high ceiling clearances with large windows provide a potential for an open style of apartment not delimited by partitioning. From our detailed surveys of occupant views in one scheme this openness was found to be enormously popular, as illustrated in Figure 3.1, which derives

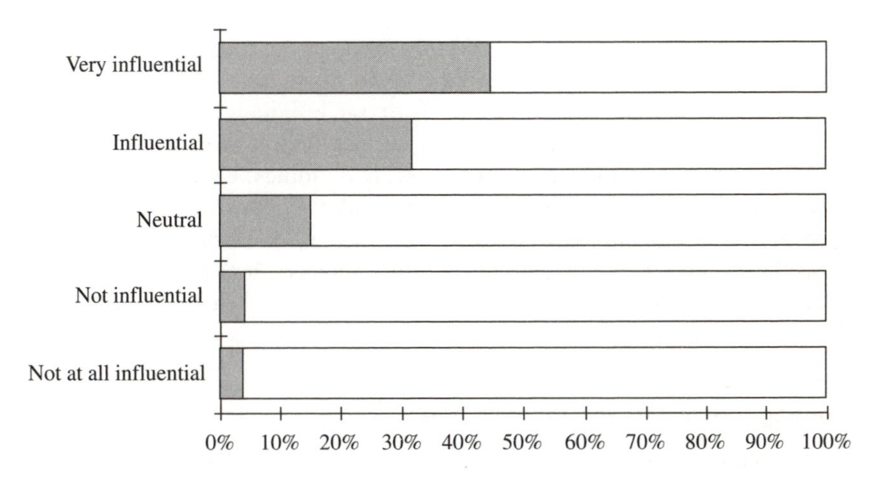

Figure 3.1 Users' perception of interior design quality as a purchase criterion.

from the survey. The enthusiasm that occupants expressed in their general comments on the survey sheets was even more striking.

Architects who are experienced in designing new spaces within conventional practices are not necessarily suited to the kind of innovation required in adaptive reuse. There was some evidence to show that where such design practices became involved they tended to resist what they found and attempted to style the building to suit their idea of what the occupant might want. This process drives up cost and extends programmes to the detriment of financial performance. In conversions of industrial buildings to office use this unwillingness to use the essential geometry of the building creatively and simply was crucial in causing commercial failure as market demand shrank in the early 1990s.

Engineering design for structures and services

Adaptive reuse of existing structures often poses very demanding and unusual problems for structural engineers in particular. Most of the problems are of a specialist technical nature and are therefore not for this set of guidelines. However, for those charged with creating a team of designers, it is evident that an engineering designer involved in this type of work needs to be comfortable with specifying exact details, working closely with contractors and using local knowledge to avoid problems. The extent of structural work varies greatly by project from virtually complete new structures to a few simple frame changes. These characteristics clearly indicate that the choice of consultant should be determined by the nature of the job and not simply by past relationships or general reputation.

The engineering of mechanical and electrical services is required to the same degree as it would be for a new building for the same use and of similar size. This was strongly indicated in the research because it was found that services were often completely stripped out and renewed during refurbishment, as indicated in Table 2.2. In most cases this meant that either complete designs were necessary for large buildings or no design was done and local service contractors simply installed conventional systems and components, particularly in smaller scale buildings. Considerable ingenuity can, however, be required to locate bulky equipment unobtrusively or to accommodate modern fixtures and fittings in traditional building fabric. This suggests that, as with structural engineers, it can be particularly beneficial to choose consultants with experience that is relevant to the type of building being refurbished.

Choosing contractors for adaptive reuse

As approximately 40% of construction industry volume in the UK is associated with refurbishment, it is no surprise to find that most contractors will, when available, tender for refurbishment work, including change of use projects. However, it can be useful to identify those who specialise in refurbishment, particularly those who have established a reputation for working to the level of specification appropriate to the project proposed. This approach equally applies when considering the particular types of construction difficulty that the project may present. The case studies done during the research provided several consistent messages in support of the above contentions, as follows.

1 Choose contractors experienced in the type of use proposed and in the level of specification intended.
2 Ensure that contractors have a good knowledge of the building technology used when the original structure was built.
3 Choose contractors who are familiar with developing the ingenious temporary structures that allow fundamental changes to be made.
4 Consider small contracts with local tradesmen where increments of work are limited and complexity is low. This particularly applies to the erection of load-bearing brick walls or installation of simple mechanical and electrical services.

In selecting contractors for this sort of work it is also important to recognise that the logistics of goods delivery and rubble removal in crowded city centres may be the prime determinant of project cost, programme and even viability. Experience in handling this issue operationally and in relation to local authorities and neighbouring property users can be invaluable.

Finally, where general contractors are involved it is important to establish a basis for working co-operatively with designers, developers and marketing people to resolve issues and achieve objectives to meet the needs of the ultimate customers who will use the space. As shown clearly in Figure 1.4, often very different perspectives exist among the participants in change of use and this is a direct concern of any general contractor charged with bringing the project team together to achieve a result.

The role of the marketeer

The research clearly showed that marketeers have a critical role to play in making change of use decisions affecting finance, design, pricing and timing. They bring a broad view of all activities in the property field and are repositories of current data on comparable prices and returns on investment. Their inclusion in the project dialogue from the outset adds value, particularly in the following areas:

1 current market trends by use type and location
2 neighbourhood characteristics and preferences

3 competitor activity
4 traditional views on occupant/client preferences (Table 3.2 illustrates this point – it is a product of the research which looked at marketeers' views on building features).

Their knowledge is central to making secure fundamental project decisions but should also be drawn upon in reviewing more detailed proposals affecting specification choices. However, it is important to realise that marketeers have no special insight on future market trends and necessarily reflect a traditionalist view of design preferences which may be at odds with real, but unstated, customer preferences. In this regard it is worth noting that little evidence was found of any market research on customer/occupant/client preferences being done by any of the firms surveyed, as indicated in Figure 3.2.

Table 3.2 Market group's view on features that positively affect value.

	Retail	*Office*	*Industrial*	*Residential*	*Other*
Building character	43%	**86%**	26%	**63%**	6%
Period features	31%	**71%**	6%	**57%**	0%
Listed building status	6%	26%	6%	34%	0%
Brick cladding	14%	31%	26%	29%	0%
Curtain wall cladding	0%	29%	17%	6%	0%
Stone cladding	9%	26%	9%	11%	0%
Floor to ceiling height	29%	**77%**	**54%**	17%	0%
Size of windows	46%	**60%**	9%	31%	0%
Other	0%	11%	6%	0%	0%

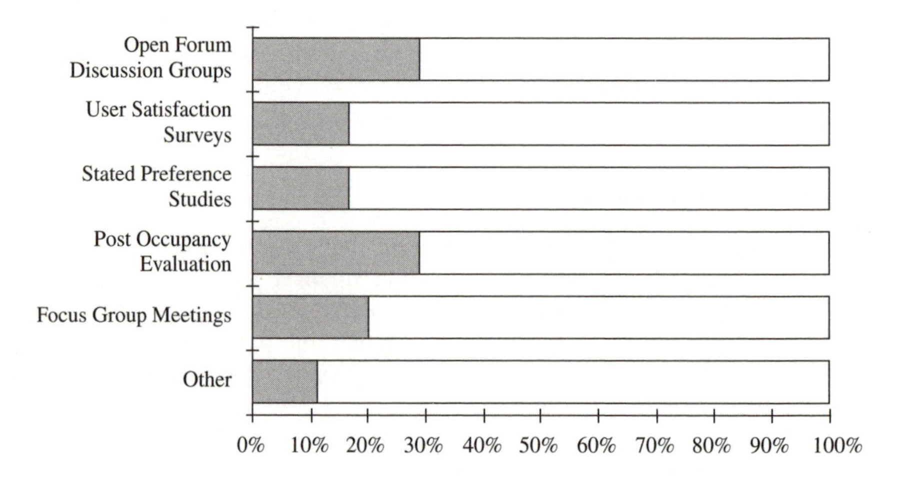

Figure 3.2 Market group techniques used to gain information about user requirements.

Including regulators in adaptation projects

Since a change of use requires full planning approval and physical changes to the building necessarily involve fire officers and district surveyors, it is inevitable that at some point these regulators will be involved in any project. The only question is, at what stage should they become involved? Experienced and successful developers and architects, in the cases studied, said that the regulators should be involved at the very earliest stages of development of proposals. Furthermore, they advised that at every stage of scheme development, regulators should be included and their advice should be sought. This is not done for altruistic reasons, but as a means of containing risk, recognising that regulators have the ability to stop a project or render it uneconomic through additional requirements.

In any event, investors always require that planning permission is obtained as a precondition of the loan. Inevitably this means that if it becomes impossible to establish a reliable dialogue with borough planners or district surveyors then the project is at risk. It is best to discover this at an early stage to avoid one of the most common causes of failure, but also, at best, to draw on the experience and judgement of planning officers to gain a better understanding of community interests. This allows others to draw on what the surveys showed (see Figure 1.4) to be the regulators' evident interest in value and robustness for further uses.

User influences

In most cases it is hardly possible to include the occupant, or user, in the adaptive reuse project. However, the research included a major survey of one user group, the owners of several hundred flats which were created in the adaptation of an large industrial building to housing. The response to the questionnaire was exceptionally good and the enthusiasm for the flats was very high despite their non-traditional design and the negative neighbourhood factors. This suggests that a lot more could be done directly by developers, marketeers and designers through market surveys, stated preference surveys, focus group reviews, etc. to understand what users want, as indicated in Figure 3.2. This may be a general issue for all involved in the built environment, but we found that it is a particular concern for adaptive reuse. The fragmented nature of the industry may contribute to this surprising lack of market research, but it is notable that even the large and long-established players are not evidently active in this area. Perhaps this represents an opportunity for newer entrants.

3.3 Decisions – managing risk and remaining robust

Key decisions and areas of risk

- Design decisions on concept or detail and achievement of programmes are more critical in adaptive reuse than in new building works.
- Phased completion can have great benefits in reducing financing costs and adjusting designs to meet demand.
- Change of use on a listed building risks loss of market value.
- Designs should be robust in allowing for the possibility of the refurbished building having uses other than that intended.

General project risk

The higher yields that are assumed by investors in change of use refurbishments (see section 3.1) relate to the higher risk associated with the project as compared with new buildings. Yet it is evident from the research that building professionals and contractors are typically more concerned about cost than about risk or value (see Figure 1.4). This represents a divergence of interests which needs to be corrected to improve any project's chance of commercial success. Thus design professionals and contracting organisations need to develop a much better understanding of the commercial risks associated with the design decisions they take. This can range from understanding of the market value of options for public area wall finishes to the risks involved in delaying the completion of the project to ensure that a material specification is unaltered just when the market is becoming time-sensitive. This complicates the work of 'producers', but if it means that their expertise is enhanced by a better understanding of the customer and the market, then they may be coming closer to the technological and commercial demands of more profitable industries and may equally be open to the greater awards available in other sectors. In this section some of the key findings of the research are discussed with the view that they may contribute to this endeavour.

Project phasing

It was evident from the case studies reviewed that many commercially successful projects benefited greatly from completing income-producing elements of work in a phased manner and that, contrarily, monolithic projects could lead to serious financial problems. The principal reasons for this are as follows.

1 The size of project loan can be reduced where unit sales (e.g. housing) are involved, thereby reducing financing costs and risk. (In one case it was discovered that the original investment in what became a landmark major scheme was less than £10,000. Revenue-producing elements of this retail development were added as surpluses allowed, and this has continued from the 1970s to today.)
2 The market can be tested and adjustments made to key factors such as unit size and cost.
3 There is an opportunity for designers to learn from user experience and introduce beneficial changes as the project develops.
4 When the total project is completed, those involved can move quickly to the next incremental project opportunity.
5 If the market collapses before the project is complete, there is a greater opportunity to put the work into abeyance without serious financial consequences.

The size and physical type of many change of use refurbishments does of course limit the scope for phasing. In such cases risk management becomes much more focused on achieving the traditional targets of time, cost and quality, as with new projects, but is also more closely connected to matters of robustness for alternative uses. This is because the larger monolithic project takes longer to complete and is therefore much more vulnerable to the ever-changing market for buildings in different use categories. Many of the most exciting adaptive reuse projects are of course monolithic, but in these cases the risks are higher and investors will be expecting exceptional rewards and seeking to spread risk through financial means rather than entirely through good project management.

Heritage listing implications

The protection of the UK's national heritage that is achieved through the listing of buildings is significant in scale, continuing to grow and has wide popular support. However, while the refurbishment of buildings is inherent to retaining them for extended or even indefinitely long periods of time, listing is not necessarily beneficial to proposals for change of use. Thus it was found in an investigation done by the RICS for English Heritage in 1993[26] (Royal Institute of Chartered Surveyors – English Heritage, 1993) that refurbishment for continued office use of listed buildings at least resulted in market values being held alongside those of unlisted buildings and in some cases slightly enhanced. However, a study by the University of

Cambridge Department of Land Economy in 1994[27] (Scanlon *et al.*, 1994) concluded that the restriction of options created by listing creates a greater degree of market uncertainty that almost inevitably reduces value, whatever the original use. It was also evident that there is a loss of robustness inherent in listing, as it is less likely that if the changed use proposed proves unviable, a quick approval for a second change could be achieved. Of course there is no possibility of demolition being approved, closing another option which is always available in unlisted change of use considerations.

The implications of these findings on the decision to pursue change of use work where listing is involved must be examined with great care. We know from the research that users in all use categories greatly value building character and historic features, but at the same time they wish to have contemporary comfort and functionality. These often competing preferences inevitably restrict the options for future use as well as limiting the choices that designers and constructors can make. While this does not rule out involvement with listed buildings, it suggests that to avoid loss of value and undue limits on change it is essential to:

1 have an open dialogue with both heritage and planning officers from the first day of the project until occupation
2 include an architectural historian or similar expert as a part-timer within the project team
3 assume a lower market value than for a similar unlisted property in business cases if direct sales are involved, in order to reflect the uncertainty referred to previously
4 conduct some form of market research before committing major funds.

The listed building is not likely to present a robust option, and there is little that can be done about this. However, there is the possibility of creating a very popular project which could serve the organisations and individuals involved well and may even prove a special case financially.

Robust decisions to protect future options

It was evident in the case studies done that, despite clear market opportunities and apparent physical practicality, the intended use for a refurbished building was not always achieved and a further refurbishment was required. It was also evident that some experienced developers aimed deliberately at mixed use as a hedge against misreading the market. Under the recent UK Planning Guidance Note No. 13 this kind of mixed use is in fact now encouraged in some circumstances by central government and is actively supported by local authorities. However, in aggregating the results of research questionnaires, as was shown in Figure 1.4, it is evident that both investors and producers generally have little apparent interest in the robustness of decisions. Marketeers, on the other hand, show a high level of interest in robustness, perhaps because they are attuned to the vagaries of the marketplace and are probably less inclined to take an optimistic view than investors, who have to believe in their judgement to justify their commitments.

However, the research results discussed in section 2.1 suggest that certain physical and locational characteristics are in themselves robust for a number of uses. **An analysis of the 77 uses, applying the comparator, can quickly identify which characteristics can and should be refurbished in such a way as to allow for a number of potential uses.** Each building will present a different set of characteristics, but a cursory examination of Table 2.4 suggests a short list of six which would merit special attention in many cases:

1 entrances
2 escape routes
3 overhead clearances
4 structural strengthening
5 subdivision of space
6 configuration of mechanical and electrical services.

Designers and contractors are seldom asked to consider these possibilities for change in their work, and if robust solutions are sought then project briefs need to introduce such considerations.

Finally, it is worth noting that regulators, particularly planners, have shown in the survey that they are not unaware of the importance of robust decisions in this area. Accordingly, it may be valuable **to include consideration of other uses either in the formal application or at least in the associated documentation**. This may well assist approval for the second use should the need arise.

3.4 Issues and opportunities in adaptive reuse

Key issues and opportunities

- The critical planning approval process depends for success on linking progress of design to stages of approval, implementing key 'development plan' points and responding to the views of influential local non-governmental bodies where possible.
- Compliance with planning policy, quality of exterior design and completeness and clarity of proposals are crucial to the success of planning applications.

- Differences of interest and perception by the key decision agents in adaptive reuse can be quite distinct and only in aggregate can be managed to create successful project teams which balance cost, value, risk and robustness.
- Discovering the attitudes and preferences of potential owner/occupiers can be a low-cost, high-value exercise. The research suggests some key housing preferences in London.

Planning approval and regulation issues

In the UK, full planning approval is required for all change of use applications, yet our questionnaire responses showed that 35% to 40% of architects submit outline planning applications at stage B or C in the RIBA plan of work, as shown in Figure 3.3. At this stage in design development full approval is not possible. This may go some way towards explaining why planning officers, when asked, said that inadequate information was the most important single reason for rejection of applications. This suggests that perhaps the quality of the dialogue between these two key players is not always of the highest order, but it also suggests that the remedy is not too difficult to find for the experienced project manager.

Generating such a dialogue may not in fact be difficult, as dialogue appears to be expected by planning officers, as suggested by their response to a being asked when they would expect to see a developer. Their responses are illustrated in Figure 3.4.

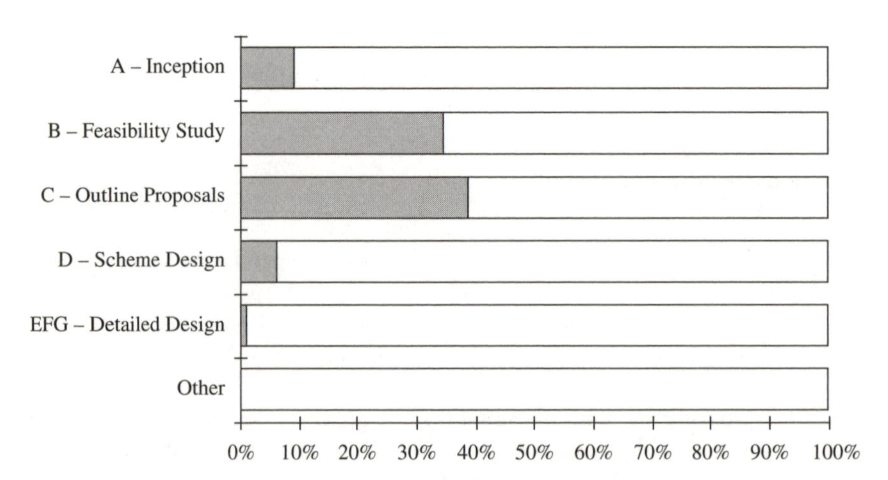

Figure 3.3 RIBA stage at which outline planning permission is sought.

In fact it is quite clear from the planning officers' response to two questions about flexibility and standards that cross a wide range of issues that they consider themselves to be open to a very wide-ranging dialogue, as indicated in Figures 3.5 and 3.6. These results suggest that the difficulty may lie with some architects who may feel that extensive dialogue may tie their hands unnecessarily. There is no evidence for this view, but there is considerable evidence of the disbenefits of not creating a positive dialogue between these two parties. Again, the project manager may need to ensure that such difficulties do not become hardened issues which put the project at risk.

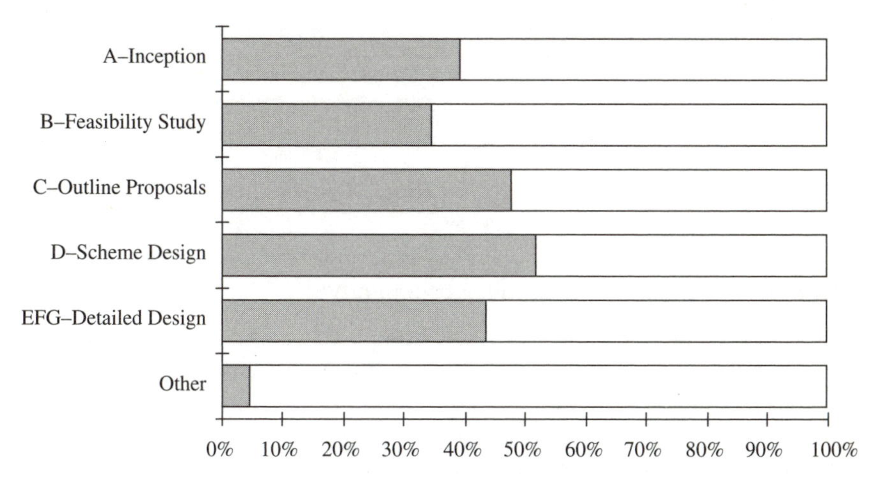

Figure 3.4 RIBA stage at which planning officers would expect to meet developers and architects.

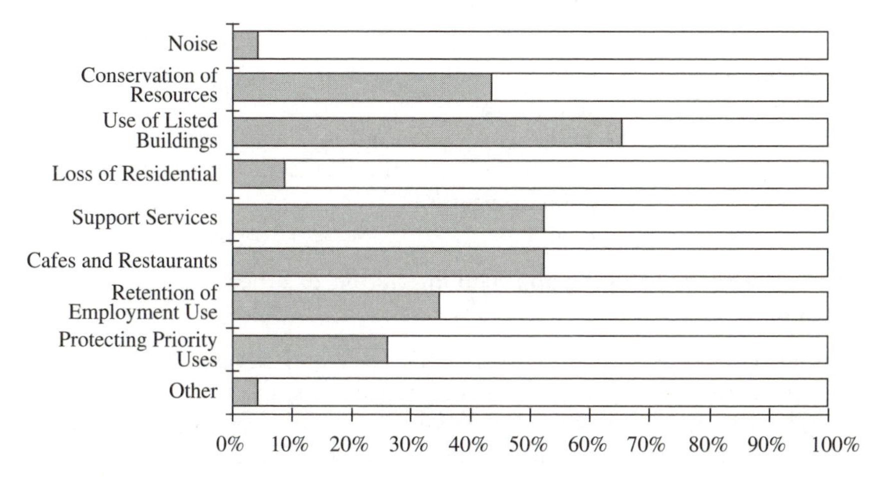

Figure 3.5 Planners' views as to which UDP policies are flexible in terms of development control.

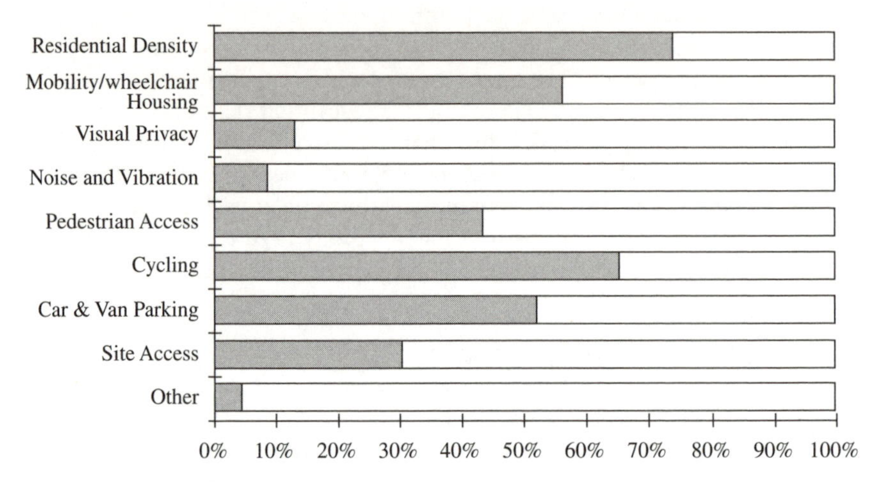

Figure 3.6 Planners' views as to which UDP standards are not flexible in terms of development control.

Planning influences of the broader community

The dialogue with planners is often complex because they are there to represent the larger community, which of course means that a wide range of social, political, religious and cultural considerations have at the very least to be given something of a hearing. In this regard it is useful to note the importance that officers place on the views of various community and other governmental groups. Figure 3.7 shows that planners are particularly tuned to organised special interests and local residents but less well tuned to commercial sector groupings. While the make-up of elected member planning committees may redress this apparent imbalance, these typical representative groups and their influence clearly need to inform those promoting adaptive reuse projects.

Notwithstanding the previous observation, it is also worth noting that 52% of planning officers indicated that the financial viability of projects was a key factor in granting approval.

The criticality of planning approval in relation to other regulations

Architects, engineers, contractors, marketeers and developers all agreed that planning approval (with Heritage approval) is the largest obstacle to the progress of a refurbishment (Figure 3.8). However, the case studies showed that producers found that the technological issues associated with concerns such as fire could be dealt with positively by the specialists on both sides of the discussion. Planning issues are much less amenable to

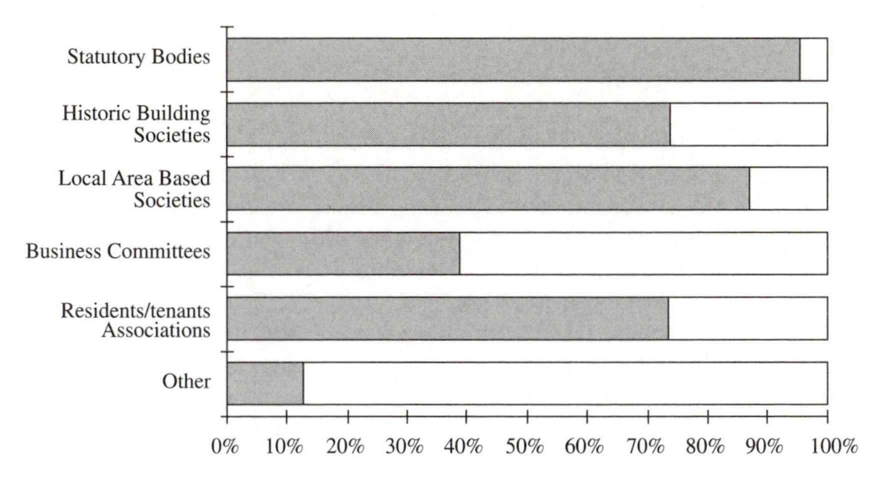

Figure 3.7 Planners' views as to which parties influence planning proposals.

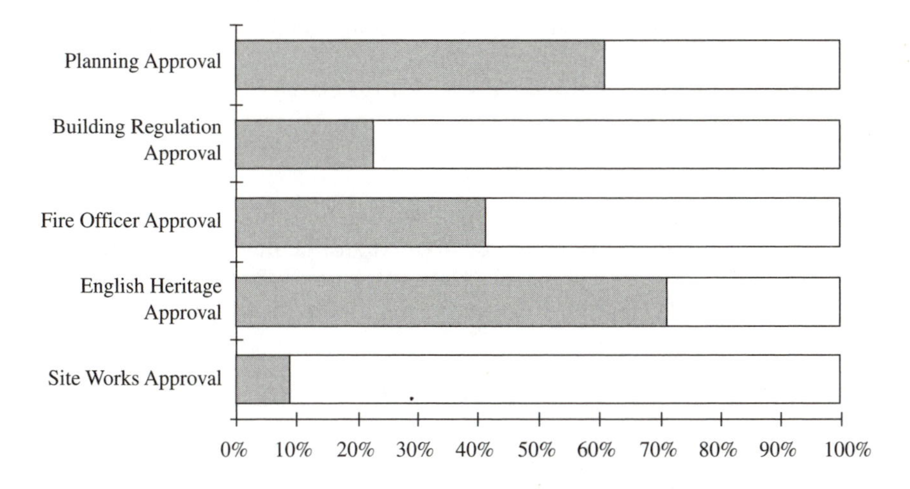

Figure 3.8 Producer group's views on regulatory approvals as a major obstacle to project progress.

this approach, as the issues are open to judgement and perception often relating to current political and social issues. There is no simple remedy to this kind of difficulty, but some appreciation of the views of planning officers and of the differing perceptions of those involved may help.

As a starting point it may be useful to note the response of the planning officers to a request asking that they identify the three key elements to a successful planning permission. Their responses were:

1 compliance with planning policy and standards
2 quality of external design
3 clarity of proposal.

The first of these was identified in some detail in Figures 3.5 and 3.6, and it can be seen that some care has to be taken here as planners' views are variable on a wide range of issues. It is therefore essential to establish a dialogue with each planning group that identifies the views they hold on a wide range of issues. All you can be sure of is that officers inevitably seek compliance, whatever the issue.

The second response deals with an even more difficult issue because of the intrinsically subjective nature of design quality. However, in many refurbishments the style of the design will be a given as the existing façade will not be changed. Emphasis will then be on the care with which restoration or small adjustments are made and the less subjective area of care in detailing will be crucial and manageable. When the façade is to be replaced or re-styled the dialogue is likely to be more general and uncertain in outcome, but good detailing is again recommended. In the latter case, choice of architect may also be crucial.

The third response is undoubtedly a plea for clarity in representation of design concepts as well as in relation to details. If there is something to hide, than designers risk failure in the approval process. If there is nothing to hide, then it is only a question of effort.

In addition to these three responses the research also looked at the contrasting causes of planning approval difficulties as seen by the 'producer group' and by planners. The results of questionnaire responses asking producers to identify issues and regulators to identify success factors

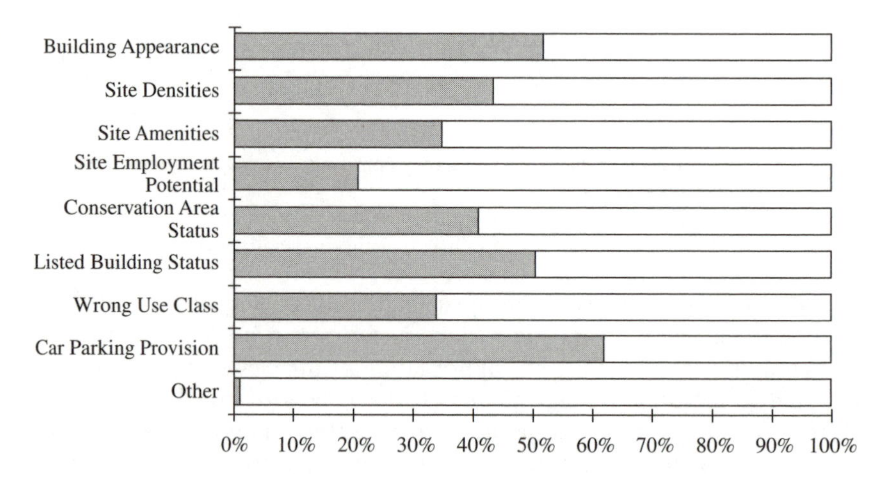

Figure 3.9 Issues perceived as planning approval difficulties by producer group.

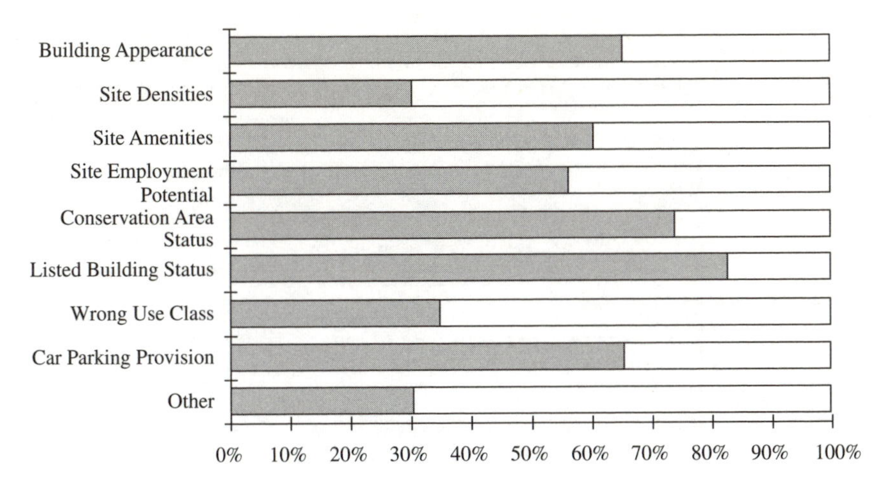

Figure 3.10 Regulator group's estimates of determining factors for a successful planning application.

show that, as seen in Figures 3.9 and 3.10, though there are a number of aspects of approval which are similarly viewed by both groups, there are significant areas of difference. For instance, building appearance is again crucial but differences in the importance given to listed building status and employment can put approval in jeopardy. The circumstances for each building are inevitably different, but it can nonetheless be instructive to consider these charts before entering the planning process.

Achieving robustness through mixed use

One of the means of achieving robustness in adaptive reuse is to seek mixed use permission from planning authorities. In this regard it is may be worth considering the views of the planning officers surveyed. These are summarised in Table 3.3. Clearly there is a concern not to mix certain uses such as industrial and residential, but the mixing of retail and industrial, for instance, is not entirely ruled out and can be a very useful mix as many light industrial products, such as in the craft-based industries, can be sold

Table 3.3 Regulator group's view on acceptable mixed uses within a single building.

	Office	*Industrial*	*Residential*	*Other*
Retail (UCO-A)	**91%**	43%	**87%**	12%
Office (UCO-B1)		**70%**	**83%**	10%
Industrial (B2–B8)			26%	7%
Residential (UCO-C)				11%

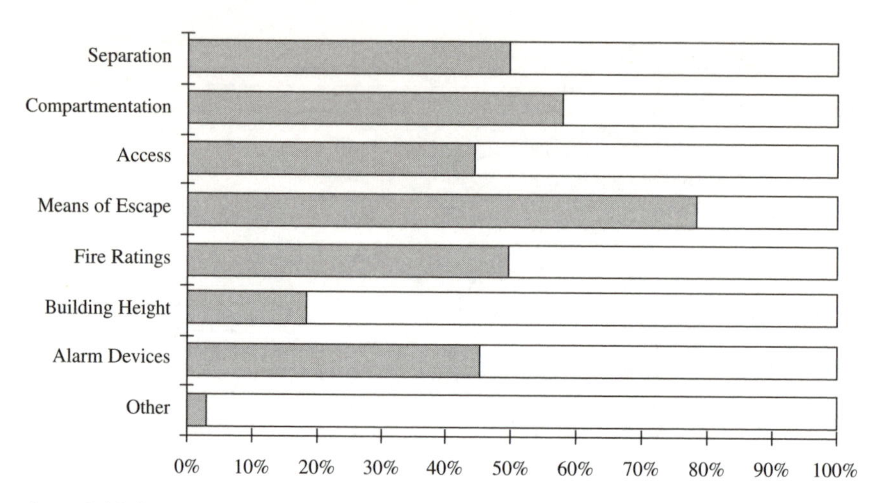

Figure 3.11 Producer group's perceived issues for difficulties with fire officer approvals.

directly from the manufacturing site. The point is that the dominance of any one use in a mixed use situation can change with relative ease or can even be allowed to result in one use only prevailing for a period.

Building regulations can complicate mixed use, however, particularly in relation to means of escape, which, as shown in Figure 3.11, 80% of designers and contractors said gave them approval difficulties even for single use situations. At times this can be an absolute barrier to creating economically viable space for two or more uses, and careful evaluation of feasibility needs to be done at the outset of projects in relation to this factor alone.

Project team characteristics

Often the major decision agents involved in the production of refurbishment schemes for change of use have markedly different views on many aspects of the work. As these differences can cause major problems (as well as unnecessary distractions), it would seem useful to all, but particularly project managers, to understand at least the different perceptions that the research identified.

To begin the commentary, it may be well to look at differing views on the selection of decision agents and employees within the team. It was found that among architects, engineers and contractors 88% use the same people for refurbishments and new builds. In apparent contrast, only 33% of developers said they used the same professional firms for new-build and refurbishment projects. Perhaps this suggests that developers need to look beyond the firm to consider the CVs of participants if they are to harvest the experience gained by individuals who have worked on other projects.

Other contrasts in views of key decision agents were seen in the responses to questions about potential project risks. Producers were seen to focus primarily on technical risks, as seen in Figure 3.12, which is probably appropriate. However, their relative lack of interest in even financial viability, let alone economic or management issues, should be cause for concern. On the other hand, marketeers take a much more balanced view of market risks, technical difficulties and management, as seen in Figure 3.13. Clearly any concerned project manager or investor could do worse than to draw producers into broader discussions than those which deal with technical

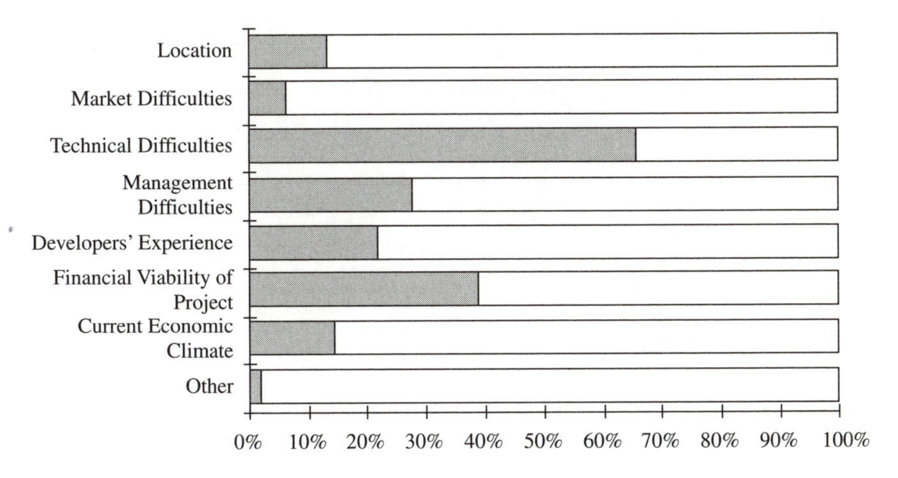

Figure 3.12 Producer group's perceptions of responsibility for project risk evaluation.

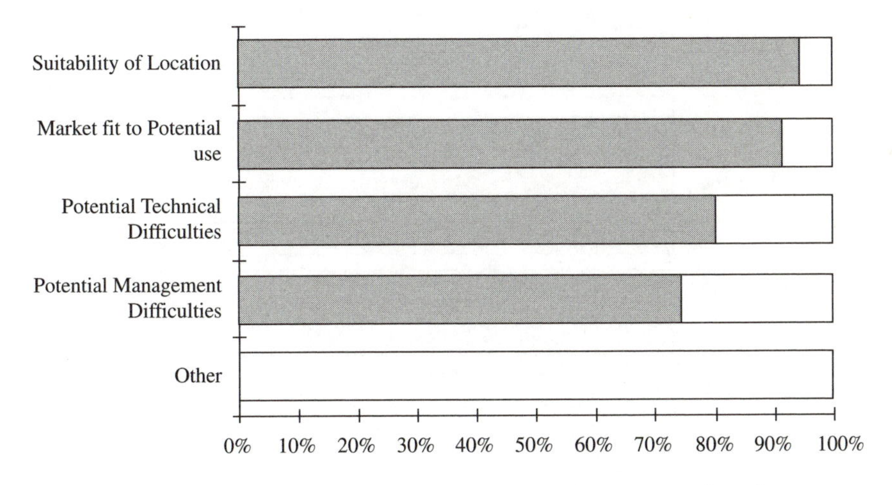

Figure 3.13 Market group's perceptions of responsibility for project risk evaluation.

issues alone if he or she is to avoid technical decisions that damage management and misread market preferences.

Another example of an important difference in perception occurs in the contrast between the producers' and the developers' approach to specification. As can be seen from Figures 3.14 and 3.15, producers typically take limited account of targeted market price in setting specifications, while developers typically take nearly as much account of this as they do of budget provision. There is also a clear difference in assumed user demand, which is the dominant criterion for developers but the sixth-ranked

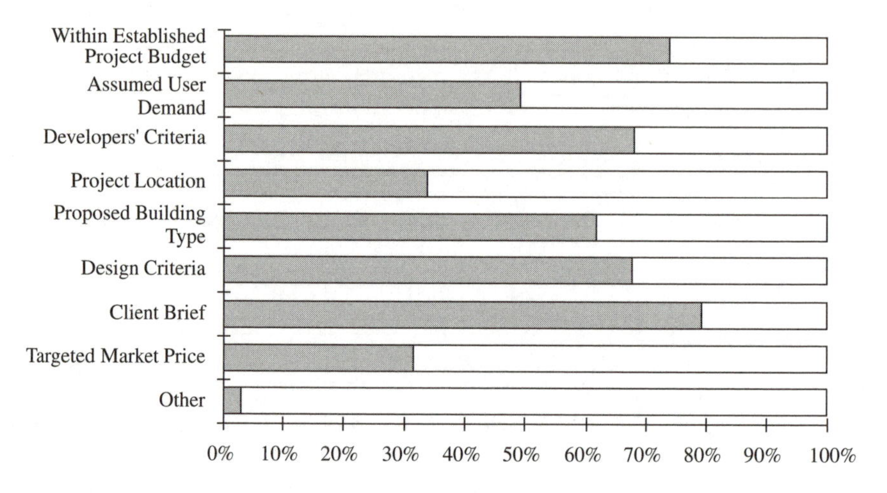

Figure 3.14 Producer group's selection of building specification.

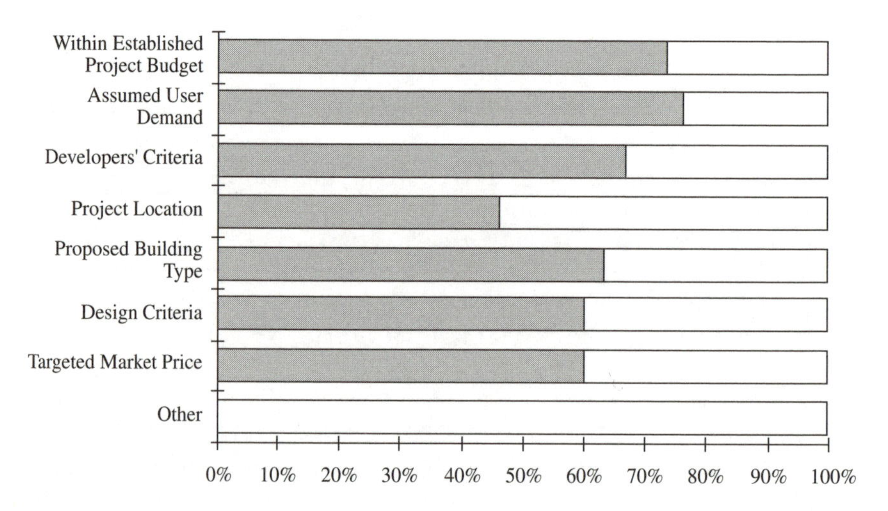

Figure 3.15 Developer group's selection of building specification.

criterion for producers. These variations are two wide to give any comfort to project managers that they can leave the specifications to the experts.

These evidently **differing views on the uniqueness of adaptive reuse, the importance of various types of risk and the consistency between specification and project objectives were seen to have been major causes of serious problems on two of the case studies considered.** Clearly, therefore, it is important to recognise these differences of interest and perception and to ensure that they do not lead to crises in the execution of the project.

User attitudes and preferences

As discussed previously, one of the findings of the research was that little if any use was made by decision agents of any method of sampling market preferences. However, in carrying out the research we used the opportunity to obtain some general information on user preferences in relation to two areas of interest.

Firstly, owner occupiers of flats which had been adapted from industrial space were asked what characteristics of the flats were most important to them. They responded as indicated in Figure 3.16.

These choices need to be interpreted in light of the style of the development surveyed. The flats provided in this scheme are warehouse style with high ceilings, large windows, open layouts and a degree of minimalism. Judging from the general comments on the questionnaires there is great enthusiasm for the development and great pleasure is taken from the 'modern' designs within a nineteenth-century industrial shell. However, the results need to be taken with some caution because the occupants were

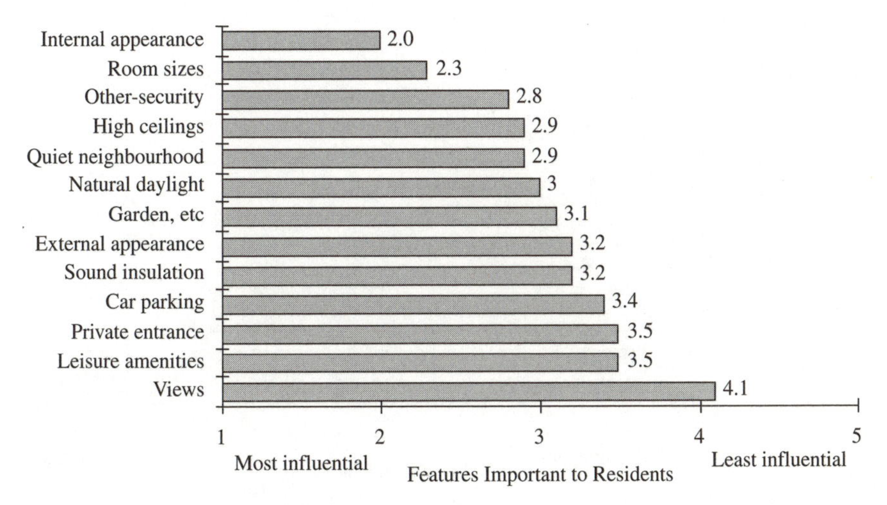

Figure 3.16 User rank of key purchase influences.

not choosing between differing potential designs or buildings but rather were in a sense explaining and justifying a choice already made.

Locational preferences related to this same development are also informative, and are indicated in Figure 3.17.

The two survey results referred to above are not illustrated here to argue that these represent the preferences of all flat buyers in London. What they do illustrate is that potential buyers for a particular style of refurbishment have quite clear views as to their preferences and these can be assessed in an orderly way. When this is set beside the normal practice of designers, contractors, marketers and developers of determining preferences by extrapolation from their individual experience, then an opportunity for beneficial change becomes apparent. Testing attitudes in a project similar to one that may be proposed can be a low-cost, high-value exercise and is recommended based on these findings.

A further step beyond the survey discussed above may be to open design to potential buyers. This is suggested by the result that showed that 95% of those surveyed said they would like to be able to choose the design of the details for their flat.

Summary

In this chapter the author has attempted to draw on the results of the research to provide fresh insights into aspects of adaptive reuse, particularly in relation to financing, obtaining planning approval, managing projects and understanding user preferences. In the final chapter the author will look to the future of adaptive reuse, with particular focus on new agendas for the sustainability of the built environment.

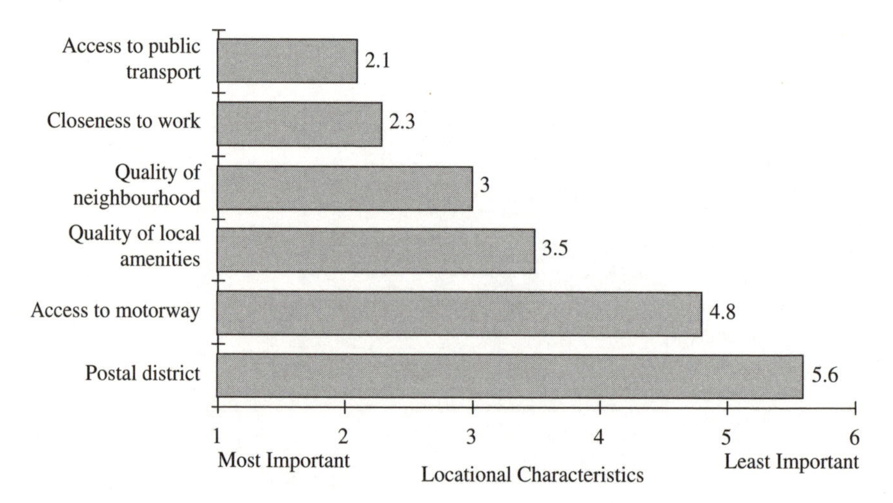

Figure 3.17 User rank of key locational characteristics.

Chapter 4

Robust buildings for changing uses

4.1 Adaptive reuse for sustainability

The final ten to fifteen years of the last century saw the beginnings of what is almost certainly a significant change in the way we use buildings and infrastructure. Following many decades of the impacts of ubiquitous electrical supplies and the telephone, which shaped much of the built environment of the twentieth century, information technology began to have a serious impact by the late 1980s. This was first visible in office buildings as networked computer terminals began to change the way we worked. Less visibly for the built environment, the same technologies also began to impact how we manufacture goods and organise distribution during the same period. These developments have themselves changed the way we work and live, but when aligned with the environmental agenda and the effect this has had on energy use and methods of production, the impacts on the built environment have been profound.

Early evidence of this change appeared as organisations in both public and private sectors found that staff were able to carry out office type work in a number of different settings through linked computers, facsimile machines and other IT communications media[28] (Becker *et al.*, 1991). These capabilities led to experiments with 'home-working'[29] (APR, Cluttons, and Gardner & Theobold, 1992) and other flexible time/space ideas. These, in turn, influenced thinking aimed at lessening environmental impacts by reducing transportation needs. From these developments, at least in part, new lifestyle preferences, characterised by looser ties between individuals and organisations, emerged. As described by Charles Handy in *The Age of Unreason*[30] (Handy, 1991) and his subsequent works, these presented organisations and individuals with profound changes. The consequent effect on the demand for office space has been both to reduce unit demand and to alter the nature of the space provided physically, locationally and tenurally.

In manufacturing, similar IT systems are used to control the robotic flexible manufacturing systems in their compact spaces largely free of all but a few technicians. An example of this was IBM's personal computer assembly space in Greenock, Scotland, which, in the author's own experience, by the early 1990s assembled all IBM PCs for Europe, the Middle East and Africa: all this in 2500 square metres of space in which fewer than ten highly skilled engineers monitored and adjusted the control computers and robots. Similar circumstances could be found (although they may well have altered by now, as change is continuous) in Vauxhall's engine manufacturing facility in the Midlands, where robots produced one complete range of engines for Europe in a space of comparable size. Similar developments can be found in healthcare, as hospitals find that the burden of post-operative care is significantly reduced by the increasing use of microsurgery techniques. Again, the effect is both to reduce and to alter the characteristics of the requirements placed on the built environment, often dramatically, from what was assumed for many years previously.

This section will explore the present and possible future extent of these changes in demand and will seek to place these in the context of the sustainability agenda. This will be done through examining how existing buildings may be adapted to meet these new needs, and how new buildings may be designed to allow sustainable adaptability to meet future needs. It seems evident from an examination of publications in this area that very little hard evidence is available on the extent and consequences of many of the technological developments discussed. Accordingly, this account will consider the direction that policy should take in order to support the development of greater sustainability of cities, in relation to both regulation and the research needed to support the continuing development of policy.

Physical aspects of sustainability

Before exploring further how to weave together aspects of the changing demand for buildings and how this interacts with the regulation and production of the built environment, a brief consideration of the physical sustainability objective and how it has so far been addressed may help to clarify how future potentials could be achieved. From a physical standpoint, sustainability of the built environment is concerned with the level at which energy transformation, material extraction and ecosystem impact can be allowed to occur in perpetuity in the creation and use of buildings and infrastructure. Questions relating to the sustainability of the activities arising from the use of this built environment and the natural environment are of course equally important, but are not addressed here. Even for this subset of the larger sustainability question, the issues are complex and difficult. A look at the publications and seminar titles of organisations such as the Green College Centre for Environmental Policy and Understanding is enough to remind one that there is much in this topic that is uncertain, much that we don't know and much that is grim in our future if we don't change our ways. In these circumstances, most policy-makers, engineers and scientists take the view that the least we can do is do the least damage in achieving worthwhile objectives to support the well-being and prosperity of our communities. This has led to a number of regulatory or formulaic prescriptions for energy use and materials application emerging in the UK concerned with items such as wall thermal conductance or environmental scoring systems for buildings[31] (Building Research Establishment, 1993). Other work has been done in the UK to design and build large-scale buildings which are naturally ventilated and use recycled materials.[32] (Happy, 1997). Those, particularly within academia, concerned with finding a way through the complexity of these issues have sought to develop or apply theoretical models to issues such as transport and patterns of growth and development. Steadman *et al.*[33], for example, applied the Tranus modelling software in looking at four potential development options for the town of Swindon in 1997.

All of these activities are of potentially great value but merely scratch the surface of the sustainability issue, simply because they affect so little of the built environment we use. This is because only 1% to 2% of the total building stock of the UK is made up of new works in a typical year[34] (Department of the Environment, 1987). The percentage of new works is even lower for roads and streets, and lower still for railways. These facts inevitably mean that we have to look at what can be done, through general refurbishment and adaptive reuse, with what we already have if we are to significantly benefit the sustainability agenda in the next 20 years. However, what we can do with existing buildings and infrastructure must, at least in part, depend on the nature and extent of the activities which the built environment serves. As pointed out above, there is gathering evidence that this demand side of the built environment equation may be significantly changing in both quantitative and qualitative ways. In any event by not examining the demand side of the equation we may be putting ourselves in the position of mere spectators who see the output of economic activity while having no understanding of what is necessary to the input. This is a poor position from which to make policy, and a near-impossible position from which to choose the particulars of the physical make-up and configuration of the built environment.

Before exploring the demand side further, it is worth noting that there are at least certain aspects of the built environment that provide reason to regard sustainability as a realistic prospect. Many structures have survived over hundreds and in some cases thousands of years; the medieval cathedrals of Europe and the aqueducts of Rome are obvious examples. As my colleague, Professor Bev Nutt has commented, sustainability is often merely a matter of planning and designing for the long term, not the short and medium as is so common. If the materials used in the first place are robust and the value placed on the maintenance of this material is appropriate, then there is good evidence that very extended life can be achieved. This may approximate sustainability, at least for a single entity, particularly if the building or structure can be adapted to a wide range of human uses, as many buildings of our historic past have been. Chapter 2 deals at length with some of the physical characteristics necessary to achieve this, at least in part. Can this longevity be extended beyond assemblies of stone, concrete and glass to the pipes, wires and fixings of the contemporary built environment? Perhaps it can, at least in the form of reusable materials, though with some expenditure of energy. Are we really back to saving string? These questions can only be looked at realistically in the context of the uses to which we put the built environment, which are examined further below.

Changing use demands

As referred to in Chapter 2, economists identify over 500 generic economic activities in the Standard Industrial Classification (SIC) system used worldwide. This forms the framework for all measurements used by government to inform economic policy and practice. Most of these activities make at least some use of buildings and infrastructure, but no attempt has been made to measure any aspect of these uses in so far as they impact the built environment except for the purpose of statutory planning. Planning activities do provide data on use, but only within the 17 much broader categories of use activity. Accordingly we are not well informed on the detail of the specific type of work that is done within buildings located variously and having many different physical characteristics. It is then perhaps not surprising that when there is a step change in how we carry out our activities within an identified field of economic activity, we cannot at the moment assess what impact this might have on the demand for space or function. Thus the working assumption of policy-makers must be that all this represents little more than adjustment of the detail within the broad sweep of economic activity. This may have worked up until now, but the sustainability agenda may oblige us to improve our information and understanding of the 'fine grain' detail of the SIC activities if we are to understand how best to use what we have and the extent to which we need to adapt, extend or demolish our buildings and infrastructure. But would the undoubtedly costly assembly of so much detail make a worthwhile contribution to policy or to physical design? A partial answer to this question may lie in looking more closely at some of the more evident examples of demand side changes. If whole sectors of economic activity are releasing buildings in substantial numbers due to demand changes, as for telephone exchanges, then perhaps a closer examination of demand detail is warranted.

Examples can be found by looking at the recent developments in organisations such as BT, IBM and Surrey County Council in the UK, all of which had specific programmes for reducing space demand and encouraging more flexibility in the work-setting choices for staff. The effects within BT emerged during the course of our research, of which BT were major sponsors. They were at the time out-looking the disposal of nearly 836,000 square metres (90 million square feet) of building floor space, while at the same time expanding their telecommunications business. They planned to retain only about 97,000 square metres. Much of the disposal (40%) was related to the obsolescence of telephone exchange buildings no longer needed to house electromechanical switching equipment because of the introduction of very much smaller computer equipment. That part of the disposal is no different from closing a factory except that the equivalent of the factory function remains. Most of the remaining disposal related to the closure of offices brought about because the number of people needed to run the business was reducing dramatically even with expected growth. However, in addition to this BT realised that with their new IT capabilities they did not need to provide one desk per person for their office-using staff. They decided to upgrade staff provision by also building some new space and in their new buildings, located often on the perimeters of cities, they have introduced office space concepts which are based on the provision of one desk on average for every 1.5 employees. This ensures that the buildings are fully used through the working day and allows for some home-working and a great deal of remote-location working.

IBM in the UK started even earlier down the path of providing less than one desk per employee. They found that when in the mid-1980s they introduced a network system linking computers on every employee's desk in every one of their 70 UK locations, the number of employees using their assigned office space at any one time declined sharply. Surveys done in 1989 and 1990 showed that this decline represented, for some groups, usages as low as 35% and often in the 50% to 60% range. All this was happening as business was growing significantly and staff numbers modestly. A programme to reduce desk provision and save space was introduced in the early 1990s under the banner of SMART (Space, Morale and Remote Technology) and led to some of the first installations of group address facilities in the UK. Professor Frank Becker of Cornell University reported on this programme and several that were similar worldwide in his report *Managing Space Efficiently*, published in 1991[35] (Becker *et al.*, 1991). Following this initial introduction, for competitive reasons, IBM's business levelled for a short period and staff reductions became inevitable. In this environment it was decided that SMART should be extended to the whole UK portfolio, and as a result space savings well beyond the percentage level of staff reductions were implemented. These reduced space provisions per employee remain in place without apparent damage to the performance of the business.

Sceptics may at this point be tempted to dismiss these cases as relating exclusively to the 'high-tech' private sector. Certainly any examination of other computer and telecommunications companies would reveal a similar picture. However, this is to ignore the much less well-publicised but widespread initiatives within the public sector, particularly local government, as reported on by Edwin O'Donnell in his 1995 UCL MSc Report[36] (O'Donnell, 1995). Working in Surrey County Council himself and leading teams involved in new ways of working, O'Donnell was able to access considerable data on the extent to which other local authorities were responding to the capabilities of new information technologies and new attitudes to work to refashion their use of their buildings and reduce the amount of space needed. It would appear from this evidence that local authorities in the UK provide a significant test bed for experiments in new ways of working. The full extent of this activity has never been measured, so it is impossible at this stage to size the effect of this, but it was notable that O'Donnell got questionnaire responses from close to 100% of those approached and most had some experience of group address initiatives.

From these illustrations it seems apparent that significant changes have already occurred in many organisations. By extension, there may be considerable scope for similar effects on both the public and private sectors as IT impacts become more ubiquitous and IT develops further. It is already evident that there is a growing belief that the impact on retailing activities may well be profound, though the jury is still out on this. If certain speculations are right, perhaps our existing city and town centres will serve as places for a type of recreational shopping, while the edge of town supermarkets increasingly become warehouses from which deliveries originate.

Before leaving the subject of demand, it is worth noting that all the above cases have the potential to affect demand for housing. Thus, as was found by BT in their studies of 'home-working', there is a gradual growth in the extent to which people choose to work at home, for at least some of their working week. The pattern of this kind of working is variable and somewhat unpredictable, but when it does occur it naturally has an effect on the built environment. At its simplest, working at home creates a need for either a separate room to cater for the work activity or at least a second purpose for an existing room. It usually increases energy use and has already led to an increase in the number of telephone lines installed in housing (report in *Financial Times*, 16 February 1999). In some ways this seems like a reversion to a pre-industrial cottage industry pattern of life, and as such could have a profound effect on what is regarded as normal use acceptable to planning authorities within residential areas. Many have found this way of working disagreeable for social reasons or impractical for their type of work, so it is unlikely to be anything but a part-time feature of life for many and a full-time feature for a few. It is, however, a real change in demand affecting housing and transport at present to an unmeasured degree.

The issues discussed above suggest that there may be much to be learned from a close look at changing use demands. This leaves the question of what can feasibly be done with the existing building stock to adapt to these demands. The following discussion considers various aspects of this adaptability potential.

Aspects of adaptability

Our research into adaptive reuse provided a useful insight into the extent and nature of these activities as they occurred in London in the mid-1990s. Examination of origin/destination figures at that time (Figure 1.2) show clearly that the dominant trend was to convert older or badly located office space into housing. This trend could also be found in a number of other cities, particularly in North America, where the author investigated this activity on visits to Toronto, Vancouver and New York in 1995. Because the cost of office space was often closer to the cost of housing in those cities, the pace and extent of the activity was greater, affecting even high-quality offices. At that time it was also evident that there was a strong trend towards change of use from industrial to housing. Resistance to change from industrial use by planners and planning committees was even then seen to be very much stronger. Thus derelict industrial space in the UK has been left for many years as determinist authorities attempted to insist on achieving employment targets for their areas[37] (APR, Cluttons, and Gardiner & Theobold, 1992). Notwithstanding this resistance some of the most significant redevelopments in London, as elsewhere, have been associated with the adaptive reuse of manufacturing, warehouse and market trading building complexes.

The research makes it clear that there are myriad choices to make in refurbishment as to materials, shapes, sizes, components and configurations for both the static and the dynamic elements of a building. To suppose then that buildings, or infrastructure of any sort, are fixed and unchanging just because they are established is to oversimplify and potentially to miss the opportunities for growth and change within what seems established. However, when 'Heritage' agendas intervene it sometimes seems that physical change should become fixed, though much can and should be changed both to visible and disguised elements to adapt to the changing needs of a living community faced with environmental challenges never previously encountered.

One of the more interesting findings of the research was that, in practice, the amount of change engendered by what was described as selective demolition was particularly significant. This kind of activity led to the repositioning of 'fixed core' items such as staircases or entranceways and often seemed to lead to totally new characteristics for a building. In the example of the telephone exchange described in Chapter 2, is was found that by demolition of the central ground floor element to create a passageway through the building the entire street could be changed by, in effect, linking it by direct pedestrian connection to both Leicester Square and Piccadilly Circus. The commercial and community benefits of this could be considerable.

Finally, reaching beyond the immediate question of reusing existing buildings, the criteria for the 77 uses identified in the comparator could be used to inform both new work and refurbishments to assist in establishing the robustness of design options. This could in itself make a contribution to the sustainability of cities as they are developed and changed. Discovering this detailed connection between design and use characteristics and extending it beyond the present level may be the first step in developing a more informed approach to how we should regulate in this area as well. However, it will be essential to allow redundancy and flexibility beyond what we can formulate in detail to ensure that the technologies of the future have at least some chance of being accommodated, as discussed below.

Adapting to changing demands to achieve sustainability

How can use be made of what is known about changing demands and the adaptive capability of buildings and infrastructure to inform our developing understanding of sustainability? Quantitative changes in demand are clearly suggested by the examples given previously, which mainly indicated that reductions in non-residential and increases in residential demand could occur, though more evidence would be needed to confirm such a trend. Qualitative changes are evident as well, but the adaptation of buildings is a mature activity and decisions can be well informed by using the techniques tested in the 'Use Comparator'. Dramatic changes such as might occur in retail could provide a lifeline to town centres and transport without heavy regulatory intervention. Such an optimistic outlook, however, ignores the history of seemingly inexorable growth of urban areas and private transport, which challenges any notion of true sustainability.

As with so many other sustainability initiatives, then, perhaps all that can be done is to incrementally improve environmental performance by avoiding those actions that lead to obviously damaging consequences and by encouraging those actions that at least minimize the damage. In this vein, a number of commentaries are offered below.

- **Redundancy:** In the first place we should respond to the uncertainty of future qualitative demands by providing some redundancy in what we build now or change. Too much floor to ceiling clearance is wasteful in both the long and the short term; too little is always wasteful in the long term though not in the short, even if uncomfortable. A generous clearance for a range of possible uses serves both the long and the short term. Similar commentary can be offered about strength, depth, height and size. Materials too should be selected to allow for these balances in benefit over the short and the long term. For both fundamental values and dimension and for materials it is always beneficial to look at the benefits that might flow from a marginal improvement and set them against short-term costs.

- **Ambiguity:** Use should be assumed to be uncertain. A variety of possible uses, not bound by the contrived categories of regulation, should be assumed likely for a building. Nothing should be done to constrain unduly the adaptation of a building to a range of future uses. Equally, a single easily defined use for a building should be avoided. Policy-makers in the UK now appear to recognise the importance of this factor.
- **Flexibility:** The most sustainable way to provide flexibility may well be to create spaces within a building which have such a quality of presence that people adapt their activities to suit the building and not the building to suit their activities. When this possibility is exhausted a building should be adaptable through its geometry, fabric and structure (in most cases) without the need to reinvent its essential morphology; nor should component cellularity constrain its ability to adapt to new technologies.
- **Constraint:** There are few physical constraints to building anything anywhere for any purpose, provided the financial resources can be found. Thus, most constraints in the built environment relate to finance and regulation. In both cases excess can damage sustainability. Too much finance leads to physical excess in size, materials and energy and must be constrained by regulation. Too much regulation leads to 'Heritage blight' as adaptation is arrested and only political intervention can redress the distortions caused. The challenge is to develop sustainability-led regulation that lays down clear objective criteria for all aspects of regulation including listing, and brings environmental performance to the forefront of design criteria.
- **Design:** It is most unlikely that sustainability can be achieved merely through the constraints of regulation. Though essential, regulation is not the basis for production of buildings, manufactured goods or food; it is a constraint on potential damage to the community by these activities. To achieve a sustainable future we need to decide what to keep, what to build, what materials to use for the buildings required to meet the requirements of users. That complex activity is largely what might be called design. Sustainability in the built environment can only be approached through informed design to meet the ever-changing needs of the users of that environment. This will have to be done for individual buildings and pieces of infrastructure as well as for the aggregation of these into the totality of towns and cities. Anything less is bound to fail the unique stringencies imposed by the sustainability agenda.

Policy options for physical sustainability

To achieve even these first steps towards sustainability will require that government play an active role in regulation of the behaviour of companies and individuals and also in expanding our knowledge of and understanding of the demand side of our use of the built environment. This will require policy initiatives in two spheres in particular.

- **Research:** A much better understanding of the activities carried out by organisations and individual users of the built environment is required. New technologies transform use both qualitatively and quantitatively, but very little money is made available by government to investigate the continuing nature and extent of this changing use. Alongside this, research is needed into the question of how different characteristics of buildings are related to the generic types of uses to which buildings are put. A new, more technologically driven approach is required to these issues if we are not to perpetuate the sterile regulatory debates driven by opinion as to which uses are appropriate to particular buildings with a given set of physical and locational characteristics.
- **Regulation:** Considerable long-term benefit would flow from a planning requirement that all permissions include a demonstration that proposals allow for a range of future uses, or at least that an appropriate range of future uses is not precluded by the design proposed. Building regulations must also move towards a set of reusability or recyclability criteria for construction works, whether related to new or refurbishment activities.

It is unlikely that these kinds of proposals, which seek to emphasise the importance of the demand side of the equation and to focus on the capability of buildings and infrastructure for adaptation, will, on their own, move us into a world which achieves sustainability. However, they are important components of the process, in the absence of which we will be unable to move successfully towards this goal.

4.2 Speculations on a more robust future

The preceding discussion has dealt largely with what we have learned from the research into adaptive reuse of buildings. In this final section the author will propose a number of speculations on possible ways we may improve our procedures, our knowledge and our regulatory regimes in the future to provide ourselves with more robust buildings better able to cope with the technological, environmental, economic and social changes that the future will surely bring. These speculations will be further to others advanced in earlier parts of the book, particularly in the previous section.

They include a number of ideas advanced by my colleagues in the research project, Professor Bev Nutt and Peter McLennan. It is hoped that the speculations might stimulate a wider debate on the work that needs to be done to at least nudge forward the development of built environment ideas in the directions being set by others who use the buildings we create.

Establishing the Use Comparator at the centre of a data system

The Use Comparator system has been established with 77 scored uses within the Bartlett School at UCL. However, it is hoped that in addition to the comparisons already tested we can regularly add buildings to the system as a means of establishing a growing database which will inform changes to the measurement scales and ratings and gradually create a greater understanding of the potential for existing buildings to be changed effectively and economically. This may well be the first step in developing the research needed to better understand use changes and their impact on buildings.

However, its primary uses are and will be to:

- identify potential user types for any given and specified redundant building
- identify potential buildings for a stated type of use.

We envisage the development of an Internet site to simplify communication, and we would hope to map the results on a currently existing advanced GIS framework.

Strategic briefing

Briefing for design is the first formal step in creating or refurbishing a building, and as such is one of the prime determinants of what we eventually experience of the built environment. To change the built environment we must then start with new approaches to the brief, taking account of the concerns outlined in the following commentaries.

- Traditionally, the briefing process first identifies the requirements of the client and user. It is assumed that an understanding of the intended use of a building provides the appropriate starting point for responsible design. For the past 20 years, this idea has formed the basis for the demand-led architectural brief and its analysis of client and user requirements[38] (RIBA, 1980).
- Today the future needs of organisations and user groups can no longer be forecast with confidence. Beyond three to five years ahead, the future is highly unpredictable.

- The traditional design brief, targeted on the current objectives of the client organisation, the contemporary requirements of the user, and the market conditions of the day, provides an inadequate basis for design and decision support systems.
- The possibilities of change to the use and purpose of a building, within a strategic approach to design, are rarely part of the briefing process.
- The research output can inform the briefs for new-build initiatives so that future buildings become more robust to functional change and have greater adaptability potential than in the past.

Development criteria

Innovative methods for the evaluation of development options, for both new-build and adaptation projects, need to be introduced. These methods must be based on composite criteria and not cost alone, to permit the systematic comparison of options in relation to:

- the probable *risk* of options
- the relative *robustness* of options
- the relative *value* of options
- the relative *utility* of options
- the relative *benefit* of options
- the relative *cost* of options.

Some of these six criteria will of course be recognised as belonging to those common to most serious attempts at financial analysis of investments, and will often be found now in evaluation methods. However, the introduction of robustness, value and benefit invite a wider dialogue than that necessary for basic funding decisions. They all look to the long term and involve a wider set of stakeholders in the decision. This can only improve the quality of decision-making in this area, which should better serve the interests of all involved.

Commentaries on technology

The avoidance of unnecessarily complex technology in electrical and mechanical service systems as well as in association with fabric and fittings appears to be an appropriate strategy for long-term robust buildings. Issues of early technological obsolescence, high cost of maintenance and inability to modify and extend with change all suggest that such services present a particular challenge to the sustainability agenda.

The assumption that higher degrees of system integration result in 'intelligent buildings' that are better able to serve user needs should be questioned. Fully integrated systems involve risks in relation to their ability to be re-differentiated and re-configured with new sub-systems extending into reconfigured building structures and spaces. Perhaps the capability for re-differentiation rather than integration is the key to improving adaptability potential in sustainable intelligent buildings.

The emerging 'flexible' working arrangements with distributed business requirements and developing cordless technologies will require fewer rather than more supporting services to be embedded within building structures and fabric. This may ease the problems of IT management and help to increase the adaptability potential of future building stock.

In conclusion

The adaptation of buildings to different uses will continue to play an important role in ensuring the continued efficient use of the building stock of communities throughout the world. The opportunity to study this activity in a large, complex city such as London provided the author and his colleagues a uniquely detailed insight into how this is done. From this has emerged a great deal of data and a number of ideas about how things might be done differently and possibly better in a few areas. We offer these ideas with what we hope is appropriate modesty, as there can be little doubt that we consistently found we were observing the work of very skilled professionals and tradesmen engaged, very expertly, in often difficult and very challenging work. Most of the observations of the author, and by extension his colleagues, are therefore aimed at the interfaces between the work of these expert practitioners, which even the most experienced are seldom able to visit. It can only be hoped that they find the result of value and that the general reader too shares this benefit.

Appendix: Use Class Framework

Common Classification	*CUC – Comparator Use Classification**	*UCO – Use Class Order*	*CI SfB Classification*
Residential	1 – Residential – individual (100.1)	C3 – Dwelling Houses (residential, home business, communal housing (<6 persons))	81 – Housing, 84 – Special housing 86 – Historical residential, 87 – Temporary, mobile residential
	2 – Residential – multiple occupancy (100.2)	C3 – Dwelling Houses	85 – Communal residential
	3 – Private housholds with employed persons (95)	C3 – Dwelling Houses	81 – Housing
Retail	4 – Retail sale in non-specialised stores, medium/small (52.1a)	A1 – Shops (shops, retail, hairdressers, undertakers, travel agencies, laundries)	34 – Trading, shops
	5 – Retail sale in non-specialised stores, large, >50 km^2 (52.1b)	A1 – Shops	34 – Trading, shops
	6 – Retail sale in non-specialised stores, high spec (medium/small) (52.1c)	A1 – Shops	34 – Trading, shops
	7 – Retail sale in specialised stores (52.2/4, 7)	A1 – Shops	34 – Trading, shops
	8 – Activities of travel agencies and tour operators; tourist assistance activities not elsewhere classified (63.3)	A1 – Shops	34 – Trading, shops
	9 – Hairdressing and other beauty treatment (93.02)	A1 – Shops	38 – Other, administration and commercial

Common Classification	*CUC – Comparator Use Classification**	*UCO – Use Class Order*	*CI SfB Classification*
Retail (cont.)	10 – Funeral and related activities (93.03)	A1 – Shops	48 – Other health, welfare facilities
	11 – Restaurants, bars, pubs, canteens (55.3/5)	A3 – Food and drink (restaurants, public houses, snack bars, hot food)	51 – Refreshment
Industrial	12 – Food and beverage (15)	B2 – General industrial	27 – Manufacturing
	13 – Tobacco products (16)	B2 – General industrial	27 – Manufacturing
	14 – Textile and textile products (17)	B2 – General industrial	27 – Manufacturing
	15 – Leather and leather clothes and products (18/19)	B2 – General industrial	27 – Manufacturing
	16 – Wood and paper products (20/21)	B2 – General industrial	27 – Manufacturing
	17 – Publishing, printing and recording (22)	B2 – General industrial	27 – Manufacturing
	18 – Machine tools and equipment (29)	B2 – General industrial	27 – Manufacturing
	19 – Electrical equipment and machinery (31/32)	B2 – General industrial	27 – Manufacturing
	20 – Motor vehicles, trailers and semi-trailers (34)	B2 – General industrial	27 – Manufacturing
	21 – Transport equipment (35.2/5)	B2 – General industrial	27 – Manufacturing
	22 – Lightweight manufacturing (36)	B2 – General industrial	27 – Manufacturing
	23 – Processed fuels and chemical products (23/24)	B4 – Special industrial	27 – Manufacturing
	24 – Basic metals (27)	B4 – Special industrial	27 – Manufacturing
	25 – Fabricated metal products (28)	B4 – Special industrial	27 – Manufacturing
	26 – Recycling of metal and non-metal waste and scrap (37)	B4 – Special industrial	27 – Manufacturing
	27 – Mineral-based products (26)	B5 – Special industrial	27 – Manufacturing
	28 – Rubber and plastic products (25)	B6 – Special industrial	27 – Manufacturing
	29 – Construction industry (45)	B8 – Storage and distribution	28 – Other industrial facilities
	30 – Wholesale trade and commission Trade (51)	B8 – Storage and distribution	28 – Other industrial facilities
	31 – Cargo handling and storage and other transport activities (63.1/2)	B8 – Storage and distribution	14 – Air transport, other transport

Common Classification	*CUC – Comparator Use Classification**	*UCO – Use Class Order*	*CI SfB Classification*
Office	32 – Gambling and betting activities (92.71)	A2 – Financial and professional services (banks, building societies, professional services, betting shops)	38 – Other, administration and commercial
	33 – Other service activities not elsewhere classified (93.05)	A2 – Financial and professional services	38 – Other, administration and commercial
	34 – Office machinery and computers (30)	B1 – Business (office, light industrial, research labs, sound and film studios)	28 – Other industrial facilities
	35 – Medical, precision and optical instruments, watches and clocks (33)	B1 – Business	28 – Other industrial facilities
	36 – Retail sale not in stores (52.6)	B1 – Business	32 – Offices
	37 – Post and courier activities (64.1)	B1 – Business	33 – Commercial
	38 – Finance, insurance and real estate industry, back office (65/70a)	B1 – Business	32 – Offices
	39 – Finance, insurance and real estate industry, principal (65/70b)	B1 – Business	32 – Offices
	40 – Computer and related activities (72)	B1 – Business	33 – Commercial
	41 – Research and development (73)	B1 – Business	33 – Commercial
	42 – General business activities and services (74)	B1 – Business	32 – Offices
	43 – Public administration and defence; compulsory social security (75)	B1 – Business	31 – Official administration, law courts
	44 – Medical practices (85.12)	B1 – Business	42 – Other medical
	45 – Other human health activities (85.13)	B1 – Business	42 – Other medical
	46 – Social work activities in accommodation (85.3)	B1 – Business	32 – Offices
	47 – Activities of membership organisations not elsewhere classified (91)	B1 – Business	53 – Social recreation, clubs
	48 – Radio and television activities (92.2)	B1 – Business	33 – Commercial
	49 – News agency activities (92.4)	B1 – Business	32 – Offices

Common Classification	*CUC – Comparator Use Classification**	*UCO – Use Class Order*	*CI SfB Classification*
Office (cont.)	50 – Extraterritorial organisations and bodies (99)	B1 – Business	32 – Offices
Other	51 – Hotels, low cost (55.1/2a)	C1 – Hotels and hostels (hotels, boarding and guest houses, old persons' homes)	88 – Other residential facilities
	52 – Hotels, standard to luxury (55.1/2b)	C1 – Hotels and hostels	88 – Other residential facilities
	53 – Higher education – residential (80.3b)	C1 – Hotels and hostels	88 – Other residential facilities
	54 – Hospital activities (85.11)	C2 – Residential institutions (residential schools and colleges, hospitals, homes (>7 (persons))	41 – Hospitals
	55 – Primary education (80.1)	D1 – Non-residential institutions (non-residential education and training, clinics, health centres, museums, public halls, places of worship, church halls)	71 – Schools
	56 – Secondary education (80.2)	D1 – Non-residential institutions	71 – Schools
	57 – Higher education – teaching (80.3a)	D1 – Non-residential institutions	72 – Universities, colleges
	58 – Adult and other education (80.4)	D1 – Non-residential institutions	72 – Universities, colleges
	59 – Veterinary activities (85.2)	D1 – Non-residential institutions	46 – Animal welfare
	60 – Activities of religious organisations (91.31)	D1 – Non-residential institutions	60 – Religious facilities
	61 – Library, archives, museums and other cultural activities (92.5)	D1 – Non-residential institutions	70 – Information facilities
	63 – Motion picture and video activities (92.1)	D2 – Assembly and leisure (cinemas, concert halls, dance and sports halls, leisure)	52 – Entertainment
	64 – Other entertainment activities (92.3)	D2 – Assembly and leisure	52 – Entertainment
	65 – Sporting activities (92.6)	D2 – Assembly and leisure	56 – Sports
	66 – Physical well-being activities (93.04)	D2 – Assembly and leisure	58 – Other recreational facilities
	67 – Building and repairing of ships and boats (35.1)	SG – Other	13 – Water transport

Common Classification	*CUC – Comparator Use Classification**	*UCO – Use Class Order*	*CI SfB Classification*
Other (cont.)	68 – Production and distribution of electricity, manufacture and distribution of gas, steam and hot water supply (40)	SG – Other	16 – Power supply, mineral supply
	69 – Collection, purification and distribution of water (41)	SG – Other	17 – Water supply, waste disposal
	70 – Sale, maintenance and repair of motor vehicles (50.1/1)	SG – Other	38 – Other commercial
	71 – Retail sale of automotive fuel (50.5)	SG – Other	18 – Other utilities
	72 – Transport via pipelines (60.3)	SG – Other	18 – Other utilities
	73 – Scheduled and non-scheduled air transport (62.1/2)	SG – Other	14 – Air transport
	74 – Telecommunications (64.2)	SG – Other	15 – Communications
	75 – Renting of machinery and equipment (71)	SG – Other	38 – Other commercial
	76 – Sewage and refuse disposal, sanitation and similar activities (90)	SG – Other	17 – Water supply, waste disposal
	77 – Washing and dry cleaning of textile and fur products (93.01)	SG – Other	38 – Other commercial
		SG – Other	37 – Protective services
		SG – Other	44 – Welfare, homes
		SG – Other	54 – Aquatic sports
		SG – Other	73 – Scientific facilities
		SG – Other	75 – Exhibition facilities
		SG – Other	90 – Common and other facilities

* Number in brackets relates to the Standard Industrial Classification (SIC).

References

1 Department of the Environment, Transport and the Regions (DETR) (1998) *Towards an Urban Renaissance*, final part of the Urban Task Force, Executive Summary, Recycling the buildings. HMSO, London.

2 Cowan, P. (1963) Studies in the growth, change and ageing of buildings. *Transactions of the Bartlett Society*, **1**, pp. 56–59.

3 Plugman, J. (2001) *Ten Shades of Green*, report of an exhibition by the Architectural League of New York at the AIA Convention, Denver, CO.

4 National Statistics, HM Government, London.

5 Department of the Environment (1987) *Re-using Redundant Buildings*. HMSO, London.

6 Department of the Environment (1991) *An Examination of the Effects of the Use Class Order 1987 and the General Development Order 1988*. HMSO, London.

7 URBED (1987) *Re-using Redundant Buildings – Good Practice in Urban Regeneration*. HMSO, London.

8 Gann, D.M. & Barlow, J. (1996) Flexibility in building use: the technical feasibility of converting redundant offices to flats. *Construction Management and Economics*, **14**, pp. 55–56.

9 McKee, W. (1996) The changing commercial property market and the relationship between landlord and tenant in the future. *Proceedings of the British Institute of Facilities Management Conference*, Cambridge, 17–18 September.

10 DEGW (Duffy, Eley, Giffone & Worthington) (1985) *Orbit 2 Executive Overview*. Harbinger Group, London.

11 As reference 4 above.

12 As reference 10 above.

13 Handy, C. (1991) *The Age of Unreason*, pp. 146–167 (Portfolios). Business Books, London.

14 Becker, F., Sims, W. & Davis, B. (1991) *Managing Space Efficiently*. Cornell University, New York.

15 Weatherall Green & Smith (1997) *International Rent Survey – Summer*. Practice Publications, London.

16 Scanlon, E., Edge, A. & Willmott, T. (1994) *The Economics of Listed Buildings*, pp. 38–40. Department of Land Economy, Cambridge University.

17 As reference 1 above.
18 Rastogi, P.N. (2001) *Managing Constant Change*. Macmillan India, Bangalore.
19 As reference 2 above.
20 Nutt, B.B. (1993) The strategic brief. *Facilities*, 2 (9), pp. 28–32.
21 As reference 8 above.
22 Sigworth, E.M. & Wilkinson, R.K. (1967) Rebuilding or renovation? *Urban Studies*, 4 (2), pp. 399–404.
23 Hillier, R.W.G. (1998) *Space is the Machine*. Cambridge University Press, Cambridge.
24 Eley, P. & Worthington, J. (1990) *Industrial Rehabilitation*, pp. 66–79. Architectural Press, London.
25 APR, Cluttons, and Gardiner & Theobald (1992) *The Home Office Report*. Publishing Business, London.
26 Royal Institute of Chartered Surveyors – English Heritage (1993) *The Investment Performance of Listed Buildings*. RICS, London.
27 As reference 16 above.
28 As reference 14 above.
29 As reference 25 above.
30 As reference 13 above.
31 Building Research Establishment (1993) *BREEAM Guidelines*. BRE, Watford.
32 Happy, J. (1997) MSC Report on Green Buildings. UCL, London.
33 Steadman, P., Holtier, S., Brown, F.E., Turner, J. & Rickaby, P.A. (1998) *An Integrated Buildings, Transport and Energy Model of Swindon*. EPSRC, Sustainable Cities Programme, Swindon.
34 As reference 5 above.
35 As reference 14 above.
36 O'Donnell, E. (1995) MSc Report on Flexible Working Practices in Local Government Offices. UCL, London.
37 As reference 25 above.
38 RIBA (1980) *The Architect's Plan of Work*. RIBA Management Handbook, London.

Index